## PRAISE FOR *WRITTEN IN B*
### Winner of the Crime Writers' Associ̶ ̶. ̶ALCS
### Gold Dagger for Nonfiction

"A fascinating overview of the human skeleton from the forensic anthropologist's point of view, with case studies that are both pertinent and entertaining."

> —Kathy Reichs, author of the Temperance Brennan *Bones* series

"Sue Black regales us with her greatest hits of forensic detective work. . . . The book is stuffed with corpses, leaking out of suitcases, stuffed in plant pots. . . . But my eye was always drawn back to her. Her extraordinarily cool, authoritative, and bloody life leaks out at the sides."

> —*Times* (London)

"A humane, wise book."

> —Crime Writers' Association judges for the
> AALCS Gold Dagger in Non-Fiction Award

"Revealing: about the human body, about the evil that men do, and—in brief, flinty asides—about herself."

> —*Sunday Times* (Scotland)

"Enjoyable . . . Black guides morbidly curious readers through baffling crime scenes, ancient crypts, and courtroom testimony to illuminate the body of evidence bones, even the smallest fragments, can offer to forensic investigators."

> —*Booklist*

"Gripping from the start, *Written in Bone* is superb—fascinating in its detail of real-life cases and written with a narrative that propels you forward to the next page."

> —Dr. Richard Shepherd, author of *Unnatural Causes*

"Sue Black shows the positive side of science, specifically forensics, delving into the practice's application in criminal investigations. She also explains how each part of the human body tells scientists about the life of the person who once occupied it."

—Publishers Weekly

"A fascinating, down-to-earth depiction of the human skeleton and how our lives can be shown from our bones. The cases she includes are also interesting and real-life examples of how this detailed knowledge can help ensure justice both for getting who dunnit as well as clearing who didn't."

—dearauthor.com

"An absorbing read . . . It is astonishing how much information even a fragment of bone may be capable of providing in the hands of a skilled forensic anthropologist. . . . The fascinating, sometimes disturbing, case examples show just how important those details can be in an investigation."

—Booked Out

## PRAISE FOR *ALL THAT REMAINS*
### Book of the Year, Saltire Literary Awards
### A CrimeReads Best True Crime Book of the Month

"With a disarming frankness . . . a multipronged approach to the topic of death, exploring it through scientific, sociological, historical, and philosophical lenses. . . . This is a perceptive study of a subject both deeply uncomfortable and uncommonly engrossing."

—Publishers Weekly

. . .

"Black's testimony to the nobility of her calling is a welcome and compassionate look at death and the mysteries that shroud it."

—Booklist

"Essential . . . An insightful, compelling, and often entertaining memoir about a life spent studying and reckoning with the dead and their secrets."

—CrimeReads, "The Best True-Crime Books of the Month"

"Dame Sue Black . . . writes vividly about her job identifying human remains, the events in her life that led her to this career, and the reality of death in all of our lives."

—*Book Riot*

"No scientist communicates better than Professor Sue Black. *All That Remains* is a unique blend of memoir and monograph that admits us into the remarkable world of forensic anthropology."

—Val McDermid, award-winning and bestselling author

"Most of us are terrified of death, but Sue Black shows us that death is in fact a wondrous process, intimately tied with life itself. Written with warmth and humanity, *All That Remains* reveals her life among the dead, who can surely count her as their best friend."

—Tess Gerritsen, internationally bestselling author

"Dame Sue Black writes about life and death with great tenderness but no nonsense, with impeccable science lucidly explained, and with moral depths humanely navigated, so that we can all feel better about the path we must all inevitably follow."

—Lee Child

"*All That Remains* provides a fascinating look at death—its causes, our attitudes toward it, the forensic scientist's way of analyzing it. A unique and thoroughly engaging book."

—Kathy Reichs, *New York Times* bestselling author

"This is one of those books that'll astound as it entertains. It's a little shivery and oh-so-fascinating. And in the end, *All that Remains* is a tale you can live with."

—*Marco Eagle* (part of the *USA Today* network)

"This fascinating look by a world-leading forensic scientist at what the dead can tell us is a real eye-opener. . . . Part meditation, part popular science, and part memoir . . . the book offers a close-up and startlingly clear view of a subject that makes most of us look away. . . . Extraordinary."

—*Sunday Times*

# WRITTEN
## *in*
# BONE

Also by Sue Black

*All That Remains*

# WRITTEN *in* BONE

## HIDDEN STORIES IN WHAT WE LEAVE BEHIND

## SUE BLACK

Arcade Publishing • New York

First North American Paperback Edition 2023

First published in Great Britain in 2020 by Doubleday, an imprint of Transworld Publishers

Arcade Publishing books may be purchased in bulk at special discounts for sales promotion, corporate gifts, fund-raising, or educational purposes. Special editions can also be created to specifications. For details, contact the Special Sales Department, Arcade Publishing, 307 West 36th Street, 11th Floor, New York, NY 10018 or arcade@skyhorsepublishing.com.

Arcade Publishing® is a registered trademark of Skyhorse Publishing, Inc.®, a Delaware corporation.

Visit our website at www.arcadepub.com.

10 9 8 7 6 5 4 3 2

Library of Congress Cataloging-in-Publication Data is available on file.
Library of Congress Control Number: 2021930902

Cover design by Erin Seaward-Hiatt
Cover illustration: © Getty Images/ilbusca

ISBN: 978-1-956763-36-2
Ebook ISBN: 978-1-951627-94-2

Printed in the United States of America

For Tom.

My whole life seeming to start and end with you.

# CONTENTS

# WRITTEN

## *in*

# BONE

# The Skeleton

*"Flesh forgets: bones remember"*

Jon Jefferson

Writer

It is not only in our brains that the memories of our lives are laid down. The adult human skeleton is made up of over two hundred bones and each has its own story to share. Some will tell it willingly to anyone who cares to ask; others guard it jealously until a deft, persistent scientific investigator cajoles them into revealing their truths. Our bones are the scaffold for our bodies and they survive long after the skin, fat, muscle and organs have dissolved back into the earth. They are designed to be robust, to hold us upright and to give us form, so it is logical that they should be the last sentinels of our mortal life to bear witness to the way we lived it.

We are used to seeing bones as dry and dead, but while we are alive, so are they. If we cut them they bleed, if we break them they hurt, and then they will try to repair themselves to regain their original shape. Throughout our existence they grow with us, adapting and changing as our lifestyle alters. The human skeleton is a living and complex organ that requires feeding and maintenance through nutrients transferred from our gut via the vast arterial network that surrounds it, with the equally complicated venous and lymphatic networks removing all the debris.

Minerals such as calcium and phosphorus, and trace elements such as fluoride, strontium, copper, iron and zinc, are modelled and remodelled continuously into our living bone structure to create its solidity and rigidity. But if bone were made up solely of inorganic materials it would be really susceptible to fracturing, so it also has an organic component, collagen, which builds in pliability. Collagen, a protein, takes its name from the Greek word for "glue," and it literally holds the mineral parts of bone together to provide us with a complex amalgam that maximizes both strength and flexibility.

We used to do an experiment in our school biology class which showed the respective functions of these two basic-level components. We would take two bones, usually rabbit thighs (often sourced from my father's shooting expeditions), and burn the first one in a furnace oven to remove the organic element. All we were left with was the mineral part of the bone, devoid of all the elastic components that hold it together: essentially, just ash. The bone would momentarily retain its form until you picked it up, whereupon it would suddenly crumble into dust.

The second bone we would place in hydrochloric acid, which leached out the mineral component. What remained was a "rubbery" bone shape, drained of all the minerals that had given it its rigidity. If you squeezed this between your fingers it felt like an eraser, and you could bend it in the middle so that each end was touching the other without it breaking. Neither component, organic nor inorganic, is on its own fit for purpose; in combination, they work together to provide us with the backbone of evolution and existence.

While bones may look quite solid, when you cut them open, you can see that they consist of two quite different types. Most of us will be aware of this from the animal bones in our cooked meat or those our dogs chew. The thick, outer shell (compact bone) has a dense, ivory-like appearance, while its more delicate inner latticework scaffolding (cancellous, or trabecular, bone) resembles honeycomb. The internal spaces are filled with bone marrow, which is a combination of fat and blood-producing cells. It is here that our red blood cells, white blood cells and platelets are made. Our bones, then, are much more than just a frame on which to hang our muscles. They are also a mineral store, a factory for blood components and the protectors of our internal organs.

With bone constantly remodelling throughout our lives, it is believed that the human skeleton essentially replaces itself every fifteen years. Some parts are replaced more quickly than others: cancellous bone reforms more often, while compact bone takes the longest. Over the years we may have many microfractures in our cancellous bone, where individual struts can break, so these need to be replaced promptly before the whole bone collapses. This continuous housekeeping of our skeleton largely goes on without affecting the original shape of the bone. However, since modifications will occur when parts are damaged, or as age alters how we replace those parts, the appearance of our skeleton does gradually change over our lifetime.

What we consume to nourish our bones is therefore vital in enabling our bodies to continue to function to their optimal capability. Bone mineral density probably reaches its peak in our fourth decade. Pregnant and breastfeeding mothers in particular draw on those resources, and, as we get older, we all do, leaving our bones increasingly depleted and our skeleton more brittle. This becomes particularly marked in postmenopausal women, when the protective action of oestrogen ceases due to the reduction of hormones in the body. As oestrogen depletes, so the floodgates open: bone mineral leaches out of the skeleton and is not replaced, and the bones become more fragile. This may lead to osteoporosis, which leaves us vulnerable to fractures, usually in the wrist, hip or spine, but they can occur in any part of the body as a result of a fall or any kind of trauma. It does not have to be excessive: a fracture can be caused by the simplest awkward movement.

It is in our interests to ensure that we lay down as high a mineral content as possible in our childhood and early adulthood. While we are growing, milk is still seen as the best source of calcium, the most important mineral for our bones. This was the rationale for supplying children with free milk at school, which began in the UK after the Second World War and continues to this day in the case of children under five attending nurseries. The other essential ingredient for healthy bones is vitamin D, which helps us to absorb the calcium and phosphorus they need. Vitamin D is provided by dairy products, eggs or oily fish, but the best source is the UVB rays in sunshine, which convert cholesterol in the skin into vitamin D. Deficiency can result in a variety of clinical conditions. It is in children that this is most evident. Babies who are permanently swaddled or young children who are kept indoors may develop disorders such as rickets, resulting in soft or brittle bones, which are most obvious in the lower limbs in the form of an inward or outward bowing of the legs.

Almost every area of our body, soft tissue and hard, can carry an echo of our experiences, our habits and our activities. We just need to know which tools to use to recover the evidence, decode it and interpret it. For example, addiction to alcohol is recorded as scars on the liver; a crystal meth habit in the teeth ("meth mouth'). A fat-heavy diet leaves its mark on the heart and blood vessels, and even on the skin, cartilage and bone, when the damage it causes results in the heart having to be accessed quickly by surgeons through the chest wall.

Many of these memories remain locked within our skeleton: a vegetarian diet is written into our bones; a healed collar bone may be a souvenir of that fall from a mountain bike. All those hours spent pumping iron in the gym are captured in increased muscle mass, and consequently in the enhanced sites of attachment of our muscles to our bones.

Perhaps these are not memories as we might normally define them, but they form an honest and reliable underscore to the soundtrack of our lives. For the most part it will never be heard, unless or until it is exposed to the scrutiny of others, perhaps through medical imaging, or if we die unexpectedly and our remains must be examined by those charged with the task of trying to figure out who we were when we were alive and what happened to us in death.

For this task we need people who have been trained to recognize the music. It may be unrealistic ever to expect to extract a complete song, but sometimes all it takes is a snatch of the melody—a bit like one of those quiz questions where you have to identify a piece of music from the introductory notes.

The forensic anthropologist's job is to try to read the bones of our skeleton as if they were a record, moving a professional stylus across them in search of the short, recognizable segments of body-based memory that form part of the song of a life, coaxing out fragments of the tune laid down there long ago. Usually this will be a life that has ended. We are interested in how it was lived and the person who lived it. We want to find the experiences recorded in the bones that will help to tell its story, and perhaps give the body back its name.

Within our discipline of forensic anthropology—the study of the human, or the remains of the human, for medico-legal purposes—there are four basic issues practitioners must address when confronted with a body, or parts of a body. Most of the time they will all be answered when the right person asks the right questions in the right way.

First of all, are the remains human?

When bones are found in unexpected circumstances, there is no point in the police setting up an investigation until this first question has been answered. Advising the police on the assumption that bones are human if they then turn out to belong to a dog, cat, pig or tortoise would be a very expensive mistake. The forensic anthropologist must be certain of the origin of the material in front of them, which means they must have knowledge and experience of the range of bones from common species likely to be encountered in the country where they are working.

As the UK is surrounded by sea, it is very common for the remains of all manner of creatures to be washed ashore on our coastline. Often these are of marine origin, so we have to know what all the different parts of a seal, a dolphin or a whale look like, alive or dead and decomposing.

We need to be familiar with the various characteristics of all of the bones found in agricultural animals such as horses, cows, pigs and sheep; in domestic pets, like dogs and cats, and wildlife—rabbits, deer, foxes and so on. While every bone in every animal is subtly different, there is a commonality to the form because it relates to function. A femur, or thigh bone, looks like a thigh bone, whether it is from a horse or a rabbit: there is just a big size difference and a bit of a variation in shape.

Between species which share a common ancestry, it can be more difficult to distinguish between their bones, for example, to tell whether a vertebra is from a sheep or a deer. There are few animal bones that should be confused with those of the human, provided the investigator has a basic knowledge of anatomy, but there are some to which even forensic anthropologists need to be alert. Human and pig ribs are very similar. The tail bones of a horse can look like human finger bones. The ones most likely to confuse us are those of species with which we share an ancestral link: other primates. This is not a problem that tends to arise very often in the UK, but one of the golden rules of forensic science is never to assume anything, and such cases are not unheard of, as we shall see.

Skeletal remains may be found on the surface of the land or underground. When bodies have been buried we need to take into account that this has been a deliberate act, and that it has usually been performed by a human. We expect humans to bury humans, but they also bury animals that are important to them, primarily pets. While people tend to bury pets where they like, often in their own gardens or woodland, we expect them to bury other people in the proper place—in a cemetery. So when we find a human above ground or buried somewhere unexpected, perhaps in a back garden or a field, there is a long set of questions to be answered about why this might be so. In short, there is an investigation to be had.

Secondly, we need to establish whether the remains are of forensic relevance.

A recently discovered body is not necessarily going to have been recently deposited, and setting up a murder investigation based on Roman remains is not likely to result in a solved case. On TV crime dramas the first question asked of a doctor, pathologist or anthropologist is always, "How long has he been dead, Doc?" This is not always easy to answer but, very crudely, if the body still has bits of flesh attached, if it is still wet with fat and if it smells bad, then it is likely to be of recent(ish) origin and so worthy of forensic investigation.

The difficulty arises when the bones are dry and all soft tissue has been lost. In different parts of the world, this stage will be reached at different times. In warmer climates, where insect activity can be voracious, a body can be reduced to a skeleton in a matter of a couple of weeks if left unburied. If it is buried, the rate of decomposition will be slower because the soil is cooler and insect activity restricted, and skeletonization may take anywhere between two weeks and ten years or more, depending on the conditions. In very cold, dry climates, the body may never completely skeletonize at all. This extensive range of possibilities does not impress the police, but the determination of the time death interval (TDI) is far from being an exact science.

Nevertheless, it is important to establish a reasonable cutoff point beyond which human remains are generally no longer considered to be of forensic interest. Of course, there will be some instances where, regardless of the passage of time, if bones come to light they may remain forensically relevant. For example, any juvenile bones found on Saddleworth Moor in the north-west of England will always be investigated as a possible link to the moors murders of the 1960s, committed by Ian Brady and Myra Hindley. Not all of the bodies of their victims have been discovered and both murderers have now taken whatever further information they might have been able to give us to their own graves.

In normal circumstances, though, if a skeleton belongs to some-one who has died more than seventy years ago, it is unlikely that any investigation would establish the circumstances of the death, still less lead to any conviction, and so technically the remains may be consid-ered archaeological. But this is a purely artificial demarcation, arrived at on the basis of the expectation of accountability in relation to a human life span. There are no scientific methodologies that can enable us to be sufficiently specific in terms of determining a TDI.

Sometimes context can help. A skeleton found buried next to a Roman coin in a known archaeological hotspot is unlikely to be of interest to the police. Neither is a skeleton uncovered by stormy weather from the sand dunes in Orkney. But they all have to be inves-tigated, just in case. A forensic anthropologist will make an early assessment and if that is not conclusive, we may send samples away for testing. Measuring the level of $C^{14}$, a radioactive isotope of carbon, which is created naturally in the atmosphere, in organic matter such as wood or bones is a method that has been used by archaeologists to date their important finds since the 1940s. The level of $C^{14}$ begins to decrease once a plant or animal dies so, basically, the older the bone is, the less $C^{14}$ will be present. As this particular radioactive isotope takes several thousand years to disintegrate completely, radiocarbon dating will only help us when remains are five hundred years old or more at the point when they are analysed and won't get us closer to modern times.

However, in the last century the human race has been the agent of disturbances in our radiocarbon levels through above-ground nuclear testing, and these have introduced manmade isotopes such as strontium-90, which has a half-life of only about thirty years. As strontium-90 did not exist before nuclear testing, if it is detected within the matrix of bones, it can only have got there during the life of the individual. So this can narrow down the date of death to within the last sixty years or so. However, self-evidently, with the passage of time, this methodology will cease to become effective. Never trust the pathologist on a TV show who says that the skeleton has been in the ground for eleven years. Utter twaddle.

Our third fundamental question is: who was this person?

If the remains have been confirmed as human and of recent origin, we need to find out who the individual was when they were alive. Our actual name is not, of course, written into our bones but they can often provide enough clues to lead to a possible identity. Once we have that, we can start to compare them with antemortem data, medical and dental records and familial biology. It is in identification that the critical scientific expertise of the forensic anthropologist is most frequently brought to bear. It is our job to extract the information held by the bones. Was this person male or female? How old were they when they died? What was their ethnic or ancestral origin? How tall were they?

The responses to these questions provide us with the four basic parameters by which every human can be categorized: sex, age, ethnicity and height. They make up a biological profile of the individual: for example, male, aged between twenty and thirty years, white, between 6 ft and 6 ft 3 ins in height. This profile automatically excludes those people reported missing who do not fit, thereby reducing the possibilities. To give an idea of scale, in a recent case, the biological profile cited above resulted in over 1,500 possible names for the police to investigate.

We ask the bones all sorts of other questions in the hope that they might answer. Did she have children? How did her arthritis affect how she walked? Where was that hip replacement done? When and how did she break that radius? Was she left- or right-handed? What size shoes did she wear? There is barely a single region of the body that cannot tell a part of our story, and the longer we live, the richer the narrative.

DNA identification has of course been a game-changer in reuniting the dead with their names. But it can only help if investigators have a source with which the DNA of the deceased can be compared. Source DNA matching requires the individual to have previously given a DNA sample that remains on record. Unless they are one of the minority who do so for occupational reasons, such as police officers, soldiers and forensic scientists, this will only have happened if they have been charged and found guilty of an offence. If the police believe they know who the person was, they can search for source DNA in their house, office or car and have it compared with that of a parent, sibling or offspring. Sometimes a relative may already be on the criminal database and a link can be made via that circuitous route. When molecular forensic science is unable to assist, forensic anthropology, and its focus on the bones, is often a last resource upon which to call. Until we have a name for the deceased, it is extremely difficult for the authorities to establish whether a crime has taken place that needs to be investigated, let alone to conclude the person's story to the satisfaction of the criminal justice system and their bereaved family.

Lastly, can we assist with the cause and manner of death?

Forensic anthropologists are scientists and, in the UK, are not generally medically qualified. Determining both manner and cause of death falls very clearly within the expertise and responsibility of the forensic pathologist. The "manner" of death might be, for example, that the victim was beaten around the head with a blunt instrument, while the "cause" of death may be blood loss. However, this is an area where the partnership between pathology and anthropology can work in harmony. Sometimes bones will tell us not only about who the person was, but what may have happened to them.

We ask different questions when dealing with the manner and cause of death. Does this child have too many old, healed injuries for them to have been caused by anything other than abuse? Did that perimortem fracture happen because this woman was trying to defend herself?

Experts learn to read different parts of the body for their own purposes. A clinician will look to the soft tissues and organs for signs of disease and a clinical pathologist may examine biopsies of tumours or categorize changes in cells to establish the nature or progression of a pathology or condition. The forensic pathologist will focus on the cause and manner of the death while the forensic toxicologist analyses body fluids, including blood, urine, vitreous humour from the eye or cerebrospinal fluid to determine if drugs or alcohol have been consumed.

With so many scientific disciplines all focusing on their own niche, sometimes with unblinking myopism, the bigger picture can often be obscured. For the clinician and the pathologist, the bones might be just something to crack open with pincers or electric saws in order to get to the organs inside. Only if there is trauma or obvious pathology will they be given more than a cursory glance. Forensic biologists are more interested in the cells that hide in the spaces within the bones than they are in the bones themselves. They will slice the bone and grind it down to a powder to get to the nuclear coding hidden in its depths. The forensic odontologist gets excited by teeth, but perhaps less so by the bones that hold them.

So the song of the skeleton may go unheard. And yet this is the most durable component of our bodies, often lasting for centuries, keeping its memories safe for a long time after the story told by the soft tissues has been lost.

If identity can be established from DNA, fingerprints or dental matching, nobody is much interested in the bones until all the other work is done and the experts have moved on to pastures new. It may be months, sometimes years, after a body is found before the forensic anthropologist enters the picture and the bones are at last called upon to give up their memories.

The scientist has no control, of course, over what they have to work with. The more recent the remains or the more complete the skeleton, the more of the story we can hope to recover, but unfortunately, human bodies are not always found intact or in good condition. The passage of time metes out its ravages on a discarded, concealed or buried corpse. Animals consume and destroy bones and the physical effects of weather, soil and chemistry conspire against retention of the song of a life lived.

The forensic anthropologist must be able to try to retrieve a part of its tune from just about anything, and to do that, we need to know what to look for and where to find it. If multiple bones tell a similar story we can have confidence in our opinion. If only a single bone is recovered we will necessarily need to be more cautious about how we interpret what it is saying to us. Unlike our fictional counterparts, we need to keep our feet on the ground and our heads out of the clouds.

Forensic anthropology is a discipline that deals in the memory of the recent, not the historical, past. It is not the same as osteoarchaeology or biological anthropology. We need to be ready to present and defend our thoughts and opinions in a courtroom as part of an adversarial legal process. Our conclusions must therefore always be underpinned by scientific rigour. We must research, test and retest our theories and be fully conversant with, and able to convey, the statistical probability of our findings. We need to understand and adhere to Part 19 of the Criminal Procedure Rules on expert evidence and to the CPS rules on disclosure, unused material and case management. We will, quite rightly, be robustly cross-examined. If our evidence is to be taken into consideration by a jury who will decide on the ultimate guilt or innocence of a defendant, we must be sound in our scientific understanding and interpretation, clear and comprehensible in our presentation and accurate in our protocols and procedures.

Perhaps forensic anthropology was once viewed as one of the easier routes into the interesting world of forensic science. It certainly exudes the kind of investigative charm that makes it irresistible to crime fiction. Not any longer. It is a profession, governed in the UK by a professional body with a royal charter. We must sit examinations and be retested every five years to remain active, competent and credible certified expert witnesses. There is no room in our business for the amateur sleuth.

This book takes you on a journey through the human body, examined through the lens of anatomy and forensic anthropology as they are applied in the real world. We will look at the body in segments, chapter by chapter, exploring how the anatomically trained forensic anthropologist might work to help to confirm the identity of the deceased and how we can assist the pathologist to determine manner and cause of death or the odontologist or radiologist to interpret findings relevant to their disciplines. We will look at the way our life experiences are written into our bones and how we can use science to unravel the story. I want to show you how using what we know of the bones allows us to piece together what can be extraordinary events—life is often more remarkable than fiction.

The forensic cases used as examples are all real ones, but in many of them I have changed names and locations out of respect for the dead and their families. Only where a case has gone to court and the press have published details of the protagonists have I included real names. The dead have a right to privacy.

# PART I

# THE HEAD
## Cranial Bones

# 1

# The Brain Box
# *Neurocranium*

*"Life's true face is the skull"*
Nikos Kazantzakis
Writer, 1883–1957

There is no more instantly recognizable image in the iconography of death than the human skull. Skulls, or their representations, have been used for ritual purposes throughout most cultures and civilizations since the earliest times. Today the skull is our preferred scary emblem for Hallowe'en, the adopted logo of heavy-metal rockers, bikers and ancient pirates, the international symbol for poisons and the favoured motif for the infamous goth T-shirt.

As *objets d'art*, the highly decorated human skulls of the Victorian era were curios made for trade, as were the infamous carved and sculpted crystal skulls that, it was claimed, originated from pre-Columbian Aztec or Mayan cultures. Many were eventually shown to be late nineteenth-century artefacts designed to entice and fleece the wealthy collector. Fake skulls have been used not only for the purposes of generating income but even to fabricate "evidence" to promote scientific theories. The 1912 Piltdown hoax was an attempt to convince the academic world that a new "missing link" had been found in the hierarchy of evolution between the ape and the human. In 1953, the humanesque skull said to have been discovered in gravel beds near Piltdown in East Sussex was exposed as a forgery when it was shown conclusively that, while the neurocranium, the "brain box" section of the skull, was that of a small modern human, the altered mandible (the lower jawbone) had come from an orangutan. Not the greatest moment in history for the image of the incorruptible British academic scientist.

The skull even became a hugely expensive piece of art when, in 2007, Damien Hirst created his iconic *For the Love of God*. The story behind the title was that his mother was always asking him, "For the love of God, what are you going to do next?" The result this time was an ostentatious platinum cast of a human skull, set with over 8,600 flawless diamonds, including a large, pear-shaped pink diamond placed in the centre of the forehead to represent the third, all-seeing, eye. The piece was tagged as a memento mori, a man-made object designed to help us reflect on the reality of our mortality, and to hint that perhaps art might succeed where life has failed: by scoring a victory over decay through the persistence of beauty. It reputedly cost around £14 million to make. To whom it was sold, or indeed whether it was ever sold at all, for its astronomical asking price of £50 million remains a mystery.

There are two aspects of this piece of Hirst's art that trouble me. The extravagant use of diamonds in such a potentially frivolous artwork is none of my business. However, the fact that the original skull was bought from a taxidermy shop in Islington should raise questions for us all about the ethics of being able to buy and sell the remains of our ancestors, irrespective of their antiquity. At one time or another, these remains were somebody's living son or daughter. If we would be offended by someone selling remains from our family vault, and most of us would be, surely we must extend the same courtesy to others? Secondly, the teeth were real: they were removed from the skull and inserted into the cast, which indicates that the integrity of the original remains was violated for the sake of art. Their disassociation bothers me. And so, on another level, does the suspicion that he got the position of some of the teeth wrong.

Perhaps the appeal of the symbolism of the skull lies in the fact that it is the most obviously human part of our remains and the core of "us" the "person": the home where we park our brains and the seat of our intellect, power, personality, senses and, some believe, even our soul. We tend to recognize people by their faces, not, for example, by their kneecaps. It is the part of a person with which we most commonly interact and it is the repository for our conscience, our intelligence and therefore our humanity and self. Our enduring fascination with skeletons and skulls probably also has a simpler source: we all possess and occupy a human body, and yet our own bones remain largely invisible to us and therefore a mystery.

When forensic anthropologists are called to assist the police with their investigations, it is understood that certain parts of a body may not be complete for perfectly explainable reasons. While most of us are issued with a full complement at birth, there are exceptions. Hands and feet, fingers and toes, for example, may never have formed, perhaps due to amniotic banding, a rare condition which can result in limbs or digits being amputated in the womb. During our lives, some of us may lose limbs through injury or have them surgically removed. And when human remains are discovered after death, some parts may be missing. Usually this will be due to scavenging animal activity but occasionally it may be because they have been deliberately removed or disposed of separately. In this, as in every aspect of our work, the forensic anthropologist must maintain an open mind and be prepared to attempt to extract as much information as possible from the smallest fragments.

While excavating a body from a lead coffin in the crypt of a church in London a few years ago, I commented to my colleague, "I can't find his left leg." She told me to look closer, because we always have two. Not in this case, however. Sir John Fraser had had his leg shot off by a cannon in the great siege of Gibraltar in 1782, so there wasn't one to be found. But one thing is certain: while we can carry on with our lives minus a limb or a finger or two, no human being has ever walked this earth without a head. Therefore, every skeleton has, or has had, a skull. And this is the bit we really want to find.

One set of remains I encountered, while working in London in my very early days as an anthropologist, presented me with a puzzle. I was contacted one morning by the police, who were looking for assistance with a "rather unusual" case. In all honesty, there is no typical case in our business. Almost every investigation has some element of abnormality or strangeness to it. The police asked if I could come down and advise them on the recovery of some skeletal remains from a garden and then examine the remains at the local mortuary.

The forensic strategy team met in one of those grey, featureless offices that are commonplace in police headquarters. Copious cups of tea are always provided and if you are lucky you might even get a bacon sandwich. The background to the case was laid out by the senior investigating officer (SIO).

A pleasant lady of mature years had walked, unannounced and in a state of some agitation, into her local police station and told the desk sergeant that if the police were to lift some patio slabs in the back garden of a nearby ground-floor property, they would find a body.

The woman was detained while a police search team was dispatched to the flat. When interviewed, she explained that some twenty years before, she used to care for the old lady who lived at this address. One day she had let herself into the flat to find her charge dead on the floor. She said that she had panicked and, not knowing what to do for the best, buried the body because she didn't want to get into trouble with the police. She told the landlord that the old lady had taken ill and been moved to a care home and then set about clearing out the property. This did not, however, explain why, as it later transpired, she continued to collect the old lady's pension for a couple of years after her death, which would in itself, you'd have thought, have been enough to attract some attention.

The flat was now occupied by another tenant, who was temporarily moved into alternative accommodation while the forensic team went to work. Through a set of sliding glass doors that led to the garden, they stepped on to a patio paved with grey concrete slabs. The slabs were easy to lift and, less than six inches below the surface, they discovered their first bone. It was at this point that the police had phoned me.

A full excavation and body recovery was undertaken and a complete set of skeletonized remains unearthed. All, that is, except for the head. When I informed the police of this absence, they asked me if I was sure. Maybe I had missed it? My indignation at the implication that perhaps I had not done my job properly, perhaps even that I wasn't able to recognize a head, was indescribable, and my response was terse. How do you miss something the size of a football? No, it had not been missed. Everything from the fourth cervical vertebra downwards was present, but the head, and the top three vertebrae, were most certainly not there.

At the mortuary I was able to confirm that the headless skeleton was that of an elderly female, who fitted the description provided by the informant down to the arthritis in her hands and feet and her hip replacement. We even found the belt she used to keep her trousers up, which had belonged to her late husband and had a distinctive military buckle. The pathologist reported that there was no specific evidence to indicate the manner or cause of death and agreed that the identity of the individual was probably not in question.

Her medical records indicated that her right hip had been replaced some years before but unfortunately no record had been kept of the implant number, which would have been a useful identifier. Her dentist told us that she wore dentures, but of course we had no head, and therefore no teeth. As she had no living relatives it was not possible to DNA-type against a family member.

Looking at the upper surface of the remaining cervical vertebrae, I was able to offer the opinion that the skull had been removed around the time of death. There was sufficient evidence of trauma and fracturing to suggest a forced separation. But we needed to find it.

When questioned about this notable omission in her confession, the detained woman eventually admitted that she couldn't bear to bury the head because the old lady was looking at her, so instead she cut it off, using the edge of the spade, she claimed, and put it in a plastic bag. She couldn't leave it behind in case anyone found it, so she had hidden it at her home and thereafter, every time she moved house, she took it with her. The next question, obviously, was where might we find the head now? The answer was that it was in her garden shed, in a plastic bag under a pile of flowerpots.

The police team was then dispatched to her garden shed. To her credit, she had at least told the truth about this. They returned to the mortuary with the skull in a supermarket carrier bag. My first job was to establish that the skull actually belonged to the body. These were the days when DNA was still in its infancy and the "fit" had to be based on anatomical articulation and whether the sex and age of the head matched that of the body. I had the skull and mandible, and the first and second cervical vertebrae, but the third was missing. Clearly this was where the dismemberment had taken place, and its absence meant we could not directly link the body anatomically to the skull. However, the anatomical features of the skull and mandible showed that they were most likely to be from an elderly female who, at the time of her death, did not have a single tooth in her head. I don't believe her dentures were ever found.

But the surprises just kept coming. For a start, it was clear that there were cut marks on the base of the skull and the second cervical vertebra. This indicated that, in addition to the spade, if indeed a spade had been involved, a sharp-bladed instrument, probably something like a meat cleaver, had been used. Even more importantly, I identified fracture patterns in the skull. There had been at least two blows to the head with a blunt instrument, perhaps the aforementioned spade, creating multiple fracture lines. The pathologist was satisfied that it was most likely death had occurred as a result of blunt-force trauma to the back of the skull, and that the head had probably been removed postmortem to conceal the manner of death. Maybe this was the real reason why the carer had taken the head with her every time she moved house.

The victim had never been listed as a missing person. She had no family to miss her. And how she came to meet such a violent end at the hands of someone who was allegedly her friend I cannot say. Whatever the circumstances, her carer was charged with murdering the old lady by hitting her, probably twice, possibly with the spade, which she may then have used to try to remove her head. When this didn't work, the carer had perhaps gone to the kitchen to find a suitable alternative. Having succeeded in separating the old lady's head from her body, she popped it in a plastic bag to take home with her, dug a hole under the patio and buried the rest of the body there.

She must have had a tremendous clean-up job on her hands before turning her attention to covering up her crime, clearing out the flat and presumably benefiting from her victim's possessions as well as her pension.

Perhaps the motive was money, and the crime was committed in cold blood. Perhaps it happened in the heat of the moment, as a result of an argument, or because the carer simply lost patience with the old lady and snapped. I was not privy to whatever explanation she may have offered. What is indisputable is that for over twenty years she seemed to have got away with murder, and yet eventually her conscience, or the increasing strain of sustaining the lie, led her to the front desk of the police station and an astounding confession. She ultimately pleaded guilty to murder, dismemberment, concealment of the remains and fraud related to claiming her victim's pension and will spend the rest of her few precious remaining days at Her Majesty's pleasure. Old age does not soften the sentences for our misdemeanours, especially if they include aggravated murder.

Most cases end up with a nickname and it was inevitable that this one was going to become known forevermore as the "head in the shed" murder. As I have often commented to crime writers, if they wrote some of the stuff that we come across in real life, nobody would ever believe them and their plots would be dismissed as ridiculously implausible.

In this instance, the bones told us not only that the head had been deliberately removed but that the old lady had been killed, not died of natural causes. But before we can read a human story in the bones, the first step is to make sure they actually are bones. Sometimes other objects can masquerade as pieces of a human skeleton and if we don't know what we are looking for, we can be fooled. Parts of the juvenile skeleton are often confused with animal bone, or even stones, because they may look like rounded little pieces of gravel. This is not generally a problem with the neurocranium, because this is usually well advanced in its development prior to birth. But confusion can occur.

A child-abuse investigation at a former children's home on Jersey, Haut de la Garenne, gained worldwide attention in 2008 after it was claimed that fragments of a juvenile skull had been found there. The presence of this "bone" was seen as damning evidence and the investigation intensified. It led to lurid speculation that children had been tortured and killed at the home and their remains concealed. However, laboratory testing undertaken to attempt to assess the age of a piece of the juvenile skull bone showed that it was not bone at all, but a fragment of wood, most likely from a coconut shell.

In the end, the police had to admit that they had no evidence that any murders had taken place at Haut de la Garenne. Of around 170 suspected bone fragments turned up at the site, only three were possibly human, and these were probably centuries old.

The absence of bodies, though, does not speak to the absence of cruelty, and the investigation did uncover a terrible catalogue of abuse at Haut de la Garenne and other children's homes on Jersey, dating back to the late 1940s. Several offenders were convicted, though many more escaped justice because they were no longer alive by the time the scandal came to light. But the time, effort and public money that was wasted on pursuing these false leads exposed the police and the forensic experts to severe criticism and threatened to jeopardize a crucial investigation.

What happened on Jersey goes to show that just because something looks like evidence of what you may be expecting to find, that does not make it so. If you are searching for the remains of children, you are not anticipating the discovery of coconuts. This is the evil of confirmation bias—the tendency to seek out something that confirms pre-set beliefs or theories, and to interpret findings through the lens of that prejudice: a tendency we all have to actively guard against. It is important that things like stones, pieces of wood and even bits of plastic (especially from fire scenes) are investigated thoroughly before conclusions are drawn because sometimes, a bone is just a coconut.

◊

If there is a tremendous longevity and richness in the cultural and emotive imagery of the human skull, the true wonder lies in the structure itself, its purpose, how it forms, how it grows and what it can tell about the life, and perhaps even the death, of the person who occupied it for a short space of time.

The bones of the adult human skull start to form towards the end of the second month of pregnancy. By the time the baby is born seven months later, virtually all the bones of the skull are recognizable, even when found in isolation from each other, providing you know what you are looking for. The growth and interconnection between the twenty-eight (or so) bones that comprise the adult skull make it one of the most complex areas of the human skeleton to try to understand, and to reconstruct from its pieces.

At birth, the baby's skull consists of nearly forty different bones, many of them measuring only a few millimetres. It is an area of the body that undergoes disproportionate growth in utero to accommodate the developing brain, but it must remain flexible if it is to be squeezed safely through Mum's ridiculously small pelvic canal. The "soft spots," or fontanelles, in a baby's head enable the bones to ride over each other during birth and allow the skull to stretch to accommodate a brain that is growing faster than the bone around it. The skull of a newborn baby can therefore sometimes appear deformed before the bones eventually settle back into position and the six fontanelles close. This starts at the age of two to three months and takes up to eighteen months to complete.

There are four primary functions the skull must fulfil from the moment of birth.

1. It must protect the very soft and fragile brain and its coverings.
2. It must have holes (foramina) to safely conduct nerves and blood vessels, and there must be external openings for the organs of special sense (eyes, ears, nose and mouth) to work optimally and allow us to interact effectively with our surrounding environment.

3.  It must provide space for sequential sets of teeth that are required for biting and chewing and it must develop the paired temporomandibular (jaw) joints that will let the teeth actively grind against each other to begin the process of food digestion.
4.  It must house the upper part of both the respiratory and alimentary tracts to facilitate breathing and the passage of masticated food respectively.

There are two basic divisions to the skull. The largest is the neuro-cranium, or vault, which is comprised of eight bones in the adult. The job of this rigid cranial chamber is almost solely to perform function number 1: to protect and support the delicate brain tissue. The smaller division of the skull is the viscerocranium, or face, which, by the time we reach adulthood, consists of a further fourteen bones. This takes care of most of functions 2 to 4. In a newborn baby, the viscerocra-nium is much smaller in relative terms, about one seventh of the vol-ume of the brain box.

A newborn baby therefore has a relatively big head (that's the main reason why birth is so tricky) and, because the eyes are a direct outgrowth from the brain, the orbits in the neonatal skull also look disproportionately large. The cartoonists and animators who created the characters for Disney and Warner Bros exaggerated these differ-ences between juvenile and adult heads to subliminally convey "good" and "bad" traits. A cute and non-threatening figure, such as Elmer Fudd, Bugs Bunny's adversary, was drawn as short and tubby with a big bald head, a small but chubby face, no chin and large round eyes: essentially paedomorphic, or childlike, in appearance. By contrast, an evil or threatening character—Jafar from *Aladdin* or Maleficent in *Sleeping Beauty*—would be tall and thin with a relatively small head, small sloping eyes, a big chin and a disproportionately long, thin face. Today's cartoons and CGI characters may be more sophisticated, but these defining features are still evident.

The reason for a baby's distinctive appearance is that the propor-tions of the skull are allied to two very different types of tissue: the brain and the teeth. As the brain develops long before the teeth, its growth requirements are more evident in the very young. The embryonic human nervous system starts as a flat sheet of tissue which then folds into a straw-like tube that runs down the centre of the body from what will be the brain end to the tail end. In the fourth week of intrauterine development, the brain end bends forward on the future brain stem and starts to swell like a balloon at the end of the straw.

Neurological expansion in the region of the future brain will continue at a rapid pace and will be at a fairly advanced stage of development before the protective bony scaffold of the neurocranium begins to consolidate around it. Brain tissue, and nervous tissue in general, sends out a signal that encourages bone to be laid down to help protect it, so it is not surprising that some of the earliest bones to develop in the human are in the skull, and in the area of the neurocranium in particular.

Patterns of growth such as the development of the sphenoid bone, right at the centre of the skull base, can help us to say whether it is from a fetus or a newborn. This bone is formed from six separate parts, two sections of a body and paired lesser and greater wings. In the fifth month of pregnancy the front of the body and the lesser wings fuse together. By the eighth month, this piece then melds with the back of the body of the bone. So at birth, the bone usually consists of three distinct parts: the fused bone consisting of the body and lesser wings, and the two separate greater wings.

All elements of the sphenoid will finally unite during the first year after birth. Being able to identify every small part of the sphenoid, fused or unfused, and understanding the pattern and sequence of age-related change, allows an anthropologist to establish the age of a child with considerable accuracy, just from this one bone. And there are many other bones in the skull that can provide similar quite specific guidance on age, which makes it a rich source of information.

If the cerebral hemispheres of the brain fail to develop, as happens, for example, in the clinical condition anencephaly, the bone is not encouraged to grow. As a result, the child may survive to birth, but will have a well-defined face with poorly constructed orbits for the eyes and a very rudimentary brain, with fundamentally no rigid box around it, giving the head the appearance of what has been described as a deflated balloon. Babies with this disorder tend not to live for more than a few hours, or days, at most. No brain, plus no brain box, results in a tragically short life.

The bones of the neurocranium form in a special membrane that surrounds the developing brain and therefore look different from other types of bone in the body. They are in the main constructed out of diploic bone, from the Greek for "double fold." This looks a bit like a sandwich, with a thin filling of more porous bone, almost aerated in appearance, between two layers of harder bone resembling ivory.

Sometimes the sandwich structure does not develop normally and areas of thinning may occur, leaving the skull vulnerable to damage. The inherited condition known as the Catlin mark often presents as two large, round holes in the parietal bones at the back of the skull, which is why it is sometimes known as the "eyes in the back of the head' condition. It was given its name by American biologist Dr William M. Goldsmith, who observed the defect in sixteen members of five generations of the Catlin family, and published his findings in 1922. Here, the bone simply does not develop, but because only small patches of it are affected, this does not appear to have a bearing on life expectancy. However, this area of the skull will be more vulnerable should the individual sustain a head injury.

The Catlin mark is very different to the acquired holes in the head that occur after trepanation—a historical and cultural activity seen in many parts of the world whereby holes are made in the skull of a (usually still conscious) patient by drilling, chiselling or scraping. The reason for this crude surgery may have been to try to cure crippling headaches or treat mental conditions (or to "release spirits," to which either type of illness may have been attributed). It was abandoned by most cultures by the end of the Middle Ages, although it was still being recorded in parts of Africa and Polynesia into the early 1900s. Without the benefit of modern-day anaesthetics, the pain must have been almost unimaginably excruciating, but there is a suggestion that the procedure could also give rise to a euphoric "high." It is incredible that people survived this brutal intervention but the fact that they did is evidenced by many skulls that show advanced healing around the incursion.

One vicious tool, dating back to the eighteenth century, shows how later procedures may have been performed. The hand brace, which looks a bit like a hand drill, boasts an end piece resembling a chisel with a spike in the middle, instantly recognizable to today's carpenters as a flat wood bit for a drill. It is no accident that the tools of the orthopaedics trade appear to emulate those of carpentry. Such are the similarities that I heard about one trainee surgeon in Wales who decided he could hone his surgical skills by spending a week on a building site as a carpenter's apprentice. Apparently, he was very accomplished.

The forensic anthropologist, then, can be faced with holes in a skull for a variety of reasons, many of which may not have played a part in the person's death, or indeed have resulted in death at all. It is relatively straightforward for an experienced professional to tell the difference between Catlin marks and trepanation holes. First of all, the position and symmetry differ: Catlin marks tend to be bilateral, to be found at the back of the parietal bones, and are usually symmetrical in size and position, whereas trepanation holes are more commonly uni-lateral and can occur anywhere on the neurocranium. The edges of the holes will be different, too. The rims of a Catlin mark are sharp, while with trepanation you can often see a saucerlike depression around the hole, where the bone has remodelled itself, providing the person survived the intrusion and healing had started. If the patient died as a result of the trepanation or soon afterwards, the marks of the tools used in the surgery will frequently still be visible as scores or grooves on the cut surface and unhealed fracture lines may be present.

Diploic bone is so distinctive that no other part of the skeleton is likely to be confused with it, and this makes it easy to recognize even when all you have to go on is a single, isolated fragment. However, other parts of the neurocranium are not always so easy to identify.

In trying to find out what had happened to one middle-aged woman who suddenly vanished from a small Scottish town, the only clue we had was the tiniest piece of something that the scene-of-crime officers, the SOCOs, thought might possibly be bone.

Mary had not been seen since putting her coat on to leave work five days before her disappearance was reported. The last thing she had said to her colleagues was that she was going home to throw her husband out on his ear because she had had enough of his lies and deceit. She had taken a phone call at work that day from their bank to say that there were some irregularities with the paperwork the couple had just signed to request a £50,000 bank loan. There certainly were "irregularities," because she had signed no such papers. Her husband had forged her signature.

Mary's husband had several failed businesses to his name and now faced mounting debts. She had reached the end of her tether with him and had often remarked to her friends that if she didn't turn up for work one day and the police came looking for her, they should tell them to dig up her back garden.

Now Mary had indeed gone missing, something to which her husband did not alert the police for five days. In an interview he stated that she had come home from work that day, they'd had a blazing row and she had stormed out. She had, he said, not been intending to return until she had calmed down. He thought she might have gone down to London to stay with one of their grown-up children. Needless to say, she had not.

A scene-of-crime team took control of the family home. They found some blood in the bathroom, which would eventually be DNA-matched to Mary, and when they put an endoscope down the U-bend of the bath, they recovered a small piece of chipped tooth enamel. This was not, of course, by any means enough to suggest that she was dead. She might have tripped going into the bathroom and cracked her chin on the side of the bath. An everyday fall could easily explain both the presence of her blood and the fragment of tooth enamel.

The SOCOs then checked out the kitchen, where they found some blood around the door of the washing machine. This, too, would be confirmed in due course as Mary's. From the filter of the machine they retrieved what they believed might be a small piece of bone. Before sending this for DNA analysis, they needed an anthropologist to have a look at it and to tell them, if possible, whether it was bone, whether it might be human and, if it was, where on the body it might have come from.

We have to be very careful about the order in which small pieces of evidence are analysed. It is important that all non-destructive forensic tests are completed first, before irreversible changes are caused in the evidence. This chip of bone, if that was what it was, was only about a centimetre long and half as wide. Testing it for DNA would entail grinding it down and all but destroying it. Trying to identify the bone anatomically was critical as a potential murder charge might hinge on it. It is quite possible to live without some bits of bone, whereas finding others outside the body would indicate that the person they belonged to is very probably dead.

The police brought the fragment to my laboratory and we all sat around the table while my colleague and I peered through magnifying lenses to try to figure out what we might be looking at. It was so delicate that we were loath to pick it up lest we damage it further. Situations like this are horribly stressful because we have to lay bare our thought processes in front of the officers in the room. What you think the bone may be when you start your deliberations is often not what you decide it is most likely to be at the end. Having to arrive at your conclusions out loud, via all the inevitable dead ends and red herrings, makes you worry what the police must think of you and your expertise.

But we must follow a rigorous process of evaluation, elimination and confirmation of identity. There is no alternative to experience and honest academic debate. Sadly, we cannot always conjure a Sherlock Holmes moment by holding up a shard of bone and exclaiming: "Aha! If I am not mistaken, Watson, this is a fragment of the left superior articular facet of the third thoracic vertebra from a woman twenty-three years of age who walked with a limp!" It is like having a single piece of a unique, 1,000-piece jigsaw puzzle. Because no two anatomical jigsaws are ever the same. Does it have edges? Can you see a pattern? Does that pattern occur in more than one place?

On this occasion, what we were clear about from the start was that the fragment was indeed bone, and that it came from the skull. It had a thin shell with a smooth outer covering and an inner surface that was slightly convoluted, with a ridge running across it. There was simply no other part of the body we could think of where this combination of characteristics would occur.

Now the puzzle became a process of positional elimination. It could not have come from the major bones of the vault, or skull cap, the upper part of the neurocranium, because this is composed only of diploic bone. As there was no diploic bone present, it had to be from the sides, the base or the face. As convolutions on the inner surface form in response to the sulci and gyri (ridges and valleys) of the cerebral hemispheres of the brain, we narrowed down our options to three possible sites: the orbital plate of the frontal bone (the roof of the eye socket), the squamous part of the temporal bone (at the side of the head above the ear) or the greater wing of the sphenoid bone (the temple area behind the eye, and in front of the ear, that you instinctively massage when you have a headache).

We decided it was too thick to be from the roof of the eye socket. The second proposition we ruled out because there is no corresponding ridge on that part of the temporal bone. No, the last and only place it could have come from was the sphenoid bone. This is a partial vault bone, without diploic structure, which shows indentations of the cerebral hemisphere on the internal surface and carries a ridge marking its junction with the frontal bone. This placement felt comfortable and defensible and we believed we had eliminated all other reasonable possibilities. Our debate had taken an hour and the police were clearly getting rather bored with us and our unintelligible anatomical babbling.

Finally, we had to determine whether it was from the right side or the left. If we were correct, it could only have come from the left, or the positioning and orientation of the ridge would have been reversed. Running close to this thin area of bone are very large blood vessels (middle meningeal vessels), and if this part of the skull was fractured, with this piece of bone extruded, then it was fairly safe to assume that Mary was no longer alive.

This, however, was a decision for the pathologist. He agreed with our conclusion, although he admitted that he couldn't comment on the identification of the bone fragment as it was beyond his anatomical knowledge. Far from being pleased by this tip of the hat to the depth of our own anatomical experience, we interpreted the pathologist's response as a warning bell. It meant we were likely to be called into court if the case went to trial because identification of the bone fragment would probably be crucial to the prosecution's case. And the procurator fiscal had confirmed that this was now a murder investigation.

The chip of bone went off for DNA testing and it was confirmed that it belonged to Mary. The husband changed his story. He said that his wife had come home and the row between them had got heated. He claimed that she had been holding a knife, because she was making sandwiches, and that he was fearful she was going to hurt him with it. He grabbed her by the hand and pushed her away, but she fell through the kitchen doorway and tumbled down a flight of steps, hitting her head on the concrete floor at the bottom. There was, he averred, blood and brain splattered everywhere. This, I might add, is not necessarily what happens when a head comes into contact with a concrete surface, and indeed no significant amount of blood had been found at the foot of the steps.

He said there was a large pool of blood coming from a wound on the left side of her head, near her ear. He realized that she was dead and so he carried her into the bathroom and laid her in the bath. He then cleaned up the house, wrapped her in plastic sheeting and put her into the boot of his car. At two o'clock the following morning, he drove off to dispose of the body. This was one part of his story the police were able to corroborate, as Mary's blood was found in the boot and his car was picked up by traffic cameras. He told them that he dumped her body into a local, fast-flowing river. To this day, she has never been found.

He had put his bloody clothes in the wash, unwittingly transferring that splinter of Mary's sphenoid bone into the machine with them. It is lucky that he did not run it at a hot temperature with a biological detergent, or we may not have been able to retrieve any DNA.

Had that been the case, it would have been much more difficult for the prosecution and forensic science to prove that the bone was Mary's. You would be forgiven for wondering who on earth else it could have belonged to, but of course, in the interests of a fair trial, our legal system requires that the burden of proof lies with the prosecution, and all the defence must do is introduce reasonable doubt.

As I'd feared, I was summoned to appear in court, where my intimate knowledge of the anatomy of the human body, even as represented by a mere chip of bone, was inevitably going to be tested robustly. The courtroom is an alien environment to a scientist. We can only answer the questions we are asked, and if the right questions are not forthcoming, it can be a gruelling and frustrating experience. In Scotland you cannot sit in the courtroom throughout the proceedings and therefore you go in cold, with no forewarning of the legal strategies in play, or of any evidence that has already been presented or is yet to be heard.

First the prosecution, in the shape of an advocate depute I had never met, began his evidence in chief by asking, on behalf of the Crown, about my credentials. I was then permitted to give my evidence and questioned on how I had reached my opinion. When you are appearing for the Crown, this is often the easiest part of the day as the prosecution has no desire or intention to challenge your testimony unless it assists their case in some way to do so. It was over in around an hour or so and much of that time was spent on satisfying the court that I was suitably qualified to give opinion evidence.

It is really important that the opinion you present in court is based solely on your specific area of experience and knowledge and that you do not stray outside your field of expertise. My evidence that day was simple. I believed the fragment to be bone, and I believed it to come from the left greater wing of the sphenoid bone. I could not comment on whether the individual to whom it belonged was still alive. I could not comment on whether the fragment belonged to Mary. I could not confirm how long the fragment had sat in the filter of the washing machine. I could not comment on how it had got there.

As judges and juries normally like to break for lunch quite promptly, I calculated that at least I would have to suffer no more than two or three hours of cross-examination by the defence. The defence QC was a man I know very well and respect enormously, but that doesn't necessarily make our court encounters a convivial affair. He is very good at his job and, although he denies it is an image he cultivates, he dresses the part and is well known for his penchant for the dramatic, with his mutton-chop sideburns and Sherlock Holmes pipe. If ever I do anything wrong and find myself up before a judge, he is the person I'd want as my lawyer.

In Scotland, you have to stand in the witness box, and I always slip off my shoes so that I feel grounded. Nobody can see I have done that. You are aware that the jury is watching you and so you adopt your best poker face. The defence QC was called to cross-examine me but remained seated, allowing an expectant hush to descend on the room. Then, in a move worthy of a television legal drama, he leaned down under his desk and lifted out a big, heavy textbook. Rising slowly to his feet, to emphasize its weight, he slammed it on the bench in front of him with a theatrical flourish and a blast of dust. It was the most recent edition of *Gray's Anatomy*, the anatomist's bible. His opening words, enunciated in his cultured Edinburgh accent, are burned into my memory: "Now, Professor, I am not doubting you for a moment . . ."

And so began an intense grilling in which I was questioned on how the bone develops in the child, how it grows, how it fractures, the soft tissue surrounding the bone and the process of differential exclusion that led me to my conclusion as to the specific anatomical position of the fragment and that it was from the left side of the skull and not the right. The prosecutor, by asking me questions that I was not qualified to answer, had cut off many of the other avenues that I am sure the defence QC would have liked to explore, such as whether the fragment could have belonged to someone else, how it might have got into the washing machine, and so on. Such are the dark arts of the legal process to which the expert witness must remain alert.

I was finished by lunchtime and on the train home thirty minutes later. I felt my credibility as an expert in the eyes of the jury had survived the cross-examination relatively intact; that I'd been able to convey my evidence in a manner they would understand, and to give them a realistic impression of the weight of my belief in the identification of the bone fragment without being overly dogmatic. And that was the end of my involvement.

After that, like everyone else, I had to watch the news and read the newspapers to find out what happened. It is a peculiar feeling to be so intimately involved in some parts of a process and yet to be excluded from so much of it. As scientists, we don't have a personal investment in any case—that would be unprofessional, not to mention detrimental to our own mental health—but you do experience something of a sense of closure when you read the outcome of a trial in the papers.

In this case, Mary's husband was found guilty of culpable homicide (roughly equivalent to manslaughter in English law) rather than murder, and sentenced to six years in prison. He was given an additional six-year sentence for perverting the course of justice by concealing the whereabouts of her body. He appealed and his sentence was reduced to nine years. In practice he served only half of that time, much of it in an open prison. I heard recently that shortly after his release he had moved down to the Blackpool area and remarried. A woman's capacity for trust and forgiveness never ceases to amaze.

Sometime after the appeal, I met the defence QC at a training workshop and good-naturedly berated him for giving me such a hard time in the witness box when my evidence was not particularly critical to the outcome of the case. The police had been able to identify that the bone fragment belonged to Mary through DNA and her husband had largely confessed to her death and the disposal of her body. We had to go to court because he stopped short of admitting to murder or culpable homicide. But of course, I appreciate that the best of defence lawyers will fight every point for their client, always drilling down into every single piece of evidence in search of a weakness in the evidence itself or in the expert's credentials, understanding or processes.

The QC's response to me, in his dry, Scottish drawl, was: "Aye. But you are so much more fun to question than the pathologists. They are much easier to trip up." And people wonder why I hate going to court.

Because the skull is three-dimensional, almost egg-like in shape, and comprised of so many different elements, each of which may have a slightly different structure, it is not surprising that there is a real art to the identification of injuries to it. When they are particularly complex, and especially when we have to try to fit together pieces of a shattered skull, it takes quite a lot of experience to work out what is what, what has happened and how.

I had been working as a forensic anthropologist at Dundee University for six years or so when I was asked by the police to review the mystifying death of a ninety-two-year-old man. The nature of the fractures to his skull, and indeed the whole question of the manner of his death, remained largely unexplained. As a new cold case team trawled through the evidence, some four years after the event, in search of a different thread to follow, they thought that perhaps anthropology could bring something fresh to their discussions.

At the police station the pathologist and I sat down with the cold case team and went over the details of the case to try to establish whether there was anything that might have been missed in the first investigation or which merited further follow-up. Much of the evidence was not challenged—until we came to the manner of death. The pathologist told us all he was prepared to say was that death had been due to multiple traumas to the head. He could not explain, though, how only one little spot of blood was found in the room where the man had died, or how he came to be lying face down on the floor with a piece of the frontal lobe of his brain on the carpet in front of him. The analysis of the brain had indicated that there was no tracking, which means that nothing had entered the skull, and yet somehow this piece of brain tissue had become detached and been expelled through a wound over his left eye.

Around the table all sorts of unlikely theories were thrown down, dissected, analysed and rejected. As the day wore on, they became more and more fantastical and we had to call a halt to our brainstorming session. It was clear that what we needed to do was to take all the crime-scene and postmortem photographs and X-rays of the body away with us, sit down somewhere quiet to examine them all in detail and think, think and think some more to see if we could develop a theoretical possibility that might explain the death and the wounds in the context of the crime-scene evidence. The body itself was no longer available to us because it had been cremated shortly after the man died. This is why comprehensive, clear and accurate photography is essential to every investigation: you have no way of knowing what evidence may be required in the future.

Colin had served in the Royal Navy in the Second World War. He had never married and had lived alone in his well-kept bungalow for forty years. He was well known and well liked, but kept himself to himself. He was very active and had been an excellent ice-skater, swimmer, walker and even water-skier until quite late in life. His neighbours reported that he went out early every morning to the local newsagent for his papers and had been seen doing so on his last day, a fact confirmed by the newsagent.

But later on, noticing that his milk had not been collected from the front doorstep, some of his neighbours went over to make sure all was well. When he didn't answer the doorbell they walked around his bungalow, looking in through the windows and calling out to him. Peering into the window to the spare bedroom at the back of the house, they saw him lying face down on the floor. The ambulance and the police were called but it was too late. Colin was dead. Initially, nobody suspected foul play. It was thought he had probably had a heart attack and had died where he had fallen. Only when the emergency services turned over his body did it become clear that an entirely different set of circumstances had led to his death and that it involved a second person.

As there was no forced entry to the bungalow it was possible that Colin knew his attacker. He had quite a bit of money saved up, which he kept at home, but that was untouched. Nothing else seemed to be missing, either, so burglary was ruled out as a motive.

At the postmortem examination, it was noted that the extent of the trauma to his skull was equivalent to someone falling from the fourth floor of a building or being subjected to a high-velocity impact in a vehicle collision. Yet it was clear that Colin had died where he fell, in the back bedroom of his bungalow. There was no significant blood spatter in the room, no disrupted furniture and no obvious weapon. It was a genuine riddle. The murder was well publicized but it seemed nobody had seen anything, nobody had heard anything and nobody could understand why anyone would do this to a defenceless old man who seemed not to have an enemy in the world. The cause of death was recorded as multiple trauma to the head and his body was released for cremation.

I pored over the photographs and images. To develop a solid theory that works from every angle you need uninterrupted time and colleagues who will question every hypothesis you propose thoroughly, to help you to construct what is most likely, after all else has been discarded as implausible or impossible. In this regard we do have something in common with Sherlock Holmes. We all follow the maxim: "When you have eliminated the impossible, whatever remains, however improbable, must be the truth."

When undertaking a fracture analysis, we first look to establish the sequence of events that might explain the pattern of fragmentation and thus the nature of the attack. Once a first crack has occurred in the bone, the force of any secondary or subsequent fracture that intersects with it will dissipate its force into the void created by the first fracture. In this way you can seriate the injuries by determining which happened first, which second and so on. It is rare (and some insist not possible) for a subsequent fracture to jump over an existing fracture and continue on to the other side, but it can happen if the force is sufficiently great.

The photographs of Colin's face taken before the postmortem began showed a sizeable opening to his skull at the inner corner of his left eye. This was wide enough for that small section of the frontal lobe of his brain to be extruded on to the carpet in front of where he lay. The problem was trying to work out how this could possibly have happened.

We knew from the neuropathologist that no object had penetrated the skull to cause the hole—the brain tissue had come out, but nothing had gone in—because of the absence of tracking marks on the brain. There was periorbital bruising around both eyes and some small grazes on his scalp, but little else. Nothing on a scale that prepared us for what we saw when we looked at the postmortem photographs taken once the scalp had been retracted and the underlying bone revealed. The fracturing was extensive. The neurocranium was in multiple fragments and fracture lines criss-crossed his skull like a spider's web.

The first thing we had to do was identify the primary fracture— the one that halted or impeded the progression of all the others. From the photographs and the X-ray images we were able to locate this at the back of his head. It had been caused by two blows, both leaving paired puncture wounds in the scalp, which had pushed the inner layer of diploic bone into the cranial cavity. The distance between the paired sharp marks appeared from the photos to be similar in both injuries, so we were content that the same weapon, something with two pointed projections, might have been used twice to hit him on the back of the head with tremendous power. Despite his advanced years, the bone at the back of Colin's skull was quite thick and it would have taken some force to pierce not only the scalp but also the robust layers of diploic bone in this area.

Reviewing the crime-scene photographs, we found some that showed a foot-operated bicycle pump, with two sharp projections at its base, lying on the floor of the spare bedroom. The distance between the projections looked consistent with the gap between the bone perforations on the skull, although as the pump had not been kept, or swabbed for blood, fingerprints or DNA at the time, we could not say for certain that this was the weapon used.

We believed this double blow set up a horizontal fracture line that ran almost from ear to ear around the back of Colin's head, which we were confident was the primary trauma site. Once you fix the first trauma in your mind, you can work forwards to the second one. In Colin's case this was probably a blow to his face which corresponded with a bruising pattern just above his left eye and across the bridge of his nose. Already disorientated by the blows to the back of his head, he had possibly now been hit quite violently in the front of his face (perhaps punched, as the bruising suggested that the attacker may have been wearing a ring).

This second trauma opened up a longitudinal fracture that extended in a relatively straight line from his eye to the back of his skull, where it terminated in the void created by the first fracture. There would probably have been very little blood visible at this stage. Although the skin was bruised, the punch was not likely to have caused the extensive open wound we could see at the corner of his eye.

It was more difficult to seriate the third trauma as it was clearly a violent event resulting in massive fragmentation of an already unstable skull. We had to look to elsewhere in his body for evidence of what might have happened to produce such extensive damage. The PM report had noted bruising to the rhomboid muscles of the left shoulder, the short muscles that link the inner border of the shoulder blade to the vertebral column. In the crime-scene photos, we could see an old mattress propped up against the wall of the spare bedroom. We thought it possible that the attacker had swung Colin by his left arm, bruising and ripping his rhomboids in the process, and that the crown of his head had collided with this mattress, which would have cushioned the impact.

It is not surprising that the SOCOs did not look for blood on the mattress as it might not have seemed to be relevant: the victim was, after all, lying on the floor with part of his brain excised. Like the bicycle pump, the mattress had not been tested and had been thrown away when the house was cleared.

The force of the swing resulting in the contact between Colin's head and the mattress might have been sufficient to rip his shoulder muscles and drive his vertebral column forwards through the hole at the base of his skull, causing the extraordinarily severe comminuted fracturing noted in the postmortem report. Not only would this shatter the base of his skull, it could explain the two radiating fractures, one on either side, which ran upwards towards the top of his head. The fracturing on the left was so extreme that it jumped both the first and the second fractures, and ultimately dissipated on the right side of his head. The internal bleeding from such an injury to the base of the skull would have been extensive as the fracturing passed through the large transverse venous sinuses and, as the pathologist confirmed, it would not have been consistent with survival. We can only hope that Colin was more or less unconscious by this time.

Incredibly, that was not the end of his nightmare. We still had to account for the hole near his eye. There was a small contusion on the side of Colin's head which seemed to match the pattern of the tread on a set of household steps standing in the room. Again, these steps were never swabbed for blood or DNA and had been destroyed with the rest of the contents of the room when the house was cleared. We believe that, after being swung against the mattress, he may have come to lie with his head on one of the steps, and that while he lay there, his head was stamped on, setting up paired horizontal fractures that ran across the front area of his head, from temple to temple.

Perhaps he was then pulled to his feet and thrown to the floor, because it is unlikely that he would have been able to get up by himself. As he finally came to rest on his front, in the position in which he would be found, the entire base of his fractured skull swung like a hinge, ripping open the already damaged skin above his left eye. As he rocked forward on his rotund tummy, the sharp, fractured edges of the skull may have acted like scissors, snipping the left frontal lobe of his brain and expelling the tissue through the hole in his face and on to the carpet in front of him.

Well, it was a theory. We interrogated and tested it from all directions, and it was anatomically plausible, if nauseatingly violent. It accounted scientifically for every single fracture, and in a credible sequence. Feeling nervous, almost sheepish, about our convoluted proposition, we returned to the next cold case review to present our theory: that the victim was hit twice on the back of the head with the bicycle pump, punched in the face, swung by the arm into the mattress against the wall and stamped on before finally being thrown to the floor. Everyone listened quietly as we went through the order of events, explaining our reasoning and detailing what, had the evidence still been available, could have been done to check whether what we were suggesting was tenable or indeed possible.

When we had finished, all eyes were on the pathologist, everyone sitting tight to see whether he would buy it. It was like waiting for a panel of judges to give you a score for your technique and performance. Eventually he nodded and said that, in the absence of any other alternatives, this was a possibility. But it remains to this day little more than supposition.

There was one interesting codicil to this sad case, after a young couple on holiday in Spain got talking in a bar to a man from the part of the world where Colin lived. As the evening wore on and the drink flowed, the man started to regale the couple with stories of the often violent activities of his former paramilitary days. When they asked him if he had ever done anything he regretted, he said that on a visit to his home town he had killed an old man, and he regretted that very much. The holidaymakers thought little of it, apparently, assuming it was just the drink talking.

But back at home, they were watching TV one night when Colin's murder was featured on the BBC's *Crimewatch*. Recognizing the location as the home town of the man they had met, they thought this was too much of a coincidence and decided they had better contact the police. They were hesitant, embarrassed even, about telling their story; the police, though, love a coincidence and they followed it up. The gentleman in question was known to them, but if it was true that he had murdered Colin, there was nothing they could do about it. He had, many years before, been given immunity from investigation and prosecution for turning Queen's evidence on some serious historical crimes.

After the programme was aired the press had a phone call—from the loyalist informant in question, who was and remains subject to death threats, he said, from the Ulster Volunteer Force. He insisted that the story he had told the couple in the Spanish bar had been misinterpreted, and that he had not been talking about Colin's death. He admitted to having been in the area at the time of the murder but denied any involvement in it.

To this day, we don't know why the killer, whether it was the informant or someone else, thought it necessary to attack this old man, let alone in such a violent way. Colin was not known to the police, nor was he someone likely to be the target of a criminal grudge, such as an ex-police or prison officer. There was no known link between him and the informant now living in Spain.

It is a pity that our evidence could not lead to an arrest but it at least provided some possible answers in what was a most perplexing case. And the work we did here—examining the evidence and trying to build a possible story that might explain the findings—is a big part of the forensic anthropologist's job. It does not necessarily mean it will result in an offender being brought to justice or even that we will ever discover if we were right. The fact that a case will sometimes remain frustratingly unresolved is something I'd had to come to terms with much earlier in my career. It might not make good television drama, but it is how things are in the real world.

◊

Once fully formed, there is little scope for the skull of an adult to change its shape. Each bone must fit snugly with its neighbour if it is to provide adequate protection. However, since growing bone is very plastic and malleable, the shape of a child's head can be altered.

Throughout history, various cultures have artificially modified the neurocranial part of the skull by moulding it in babies before the bones have "set," either in response to some belief that a particular shape had a beneficial effect on a person's thought processes or simply because it was more aesthetically pleasing. Among the higher social strata of certain tribes, such cranial deformation was seen for the rest of a child's life as a badge of their social status.

These alterations to the appearance of the skull were achieved by binding the baby's head between wooden boards, or winding it with tight cloths or bandages, to produce the desired shape: sometimes elongated, sometimes conical, sometimes round. It usually began about a month after birth and continued for six months, perhaps even a year or two, until the fontanelles had closed and the deformation was largely irreversible. It is claimed that there was no neurological impact on the child, but I am sceptical.

This procedure was practised across many disparate geographical locations and chronological periods, from the north and south Americas, to Iraq, Egypt, Africa, Russia and pockets of Europe and Scandinavia. In some cases, such as the "Toulousian Deformation," which persisted in rural southern France until the early twentieth century, the purpose of the binding was simply to pad and protect the delicate skull and the deformity was merely a by-product of the tradition rather than its intended outcome.

Whatever the shape of the skull, the vault can tell us a lot about an individual, particularly in relation to their sex, age and sometimes their ethnicity. Sex determination is usually associated with enlarged sites of muscle attachment in the male and a more gracile, less robust appearance in the female. There are few muscles that attach to the neurocranium, but if you palpate quite deeply into the midline space where the muscles of the back of the neck meet the bone at the base of the skull, you will probably feel a large lump of bone in the male, but not in the female. This is called the external occipital protuberance and it is generally more developed in the male as it is where a very strong ligament of the vertebral column—the ligamentum nuchae—attaches. This ligament helps to hold the vertebrae of the neck in alignment and the head balanced on the first cervical vertebra.

An Australian university recently hit the headlines when they announced, on the strength of the results of a small study, that modern-day adolescents and young adults show enhanced growth in the region of the external occipital protuberance. The fact that the study was based on a sample size of only 218 did not deter them from opining that this was caused by young people adopting a "heads down" posture consistent with prolonged use of electronic devices. The Bronze

Age skeletons I studied for my honours degree frequently had very well-developed occipital spurs, but no matter how hard I searched, I never found their mobile phones.

Science may be wonderful but pseudoscience can be dangerous. It is very tempting to share our theories with the world but we must be careful not to extrapolate over-enthusiastically from limited observations. We cannot mislead an investigation or a courtroom with unproven information based on our own fanciful hobby horse.

The little lumps of bone at the back of your ears, called the mastoid processes (from the Latin meaning "little breasts"), can also be a useful, though not infallible, indicator of sex. These are the site of attachment for the long sternocleidomastoid muscle that runs between the front of your chest and the back of your ear. You can see this muscle if you stretch your neck and twist your head to one side. The stronger the muscle, the larger the lump of bone, so mastoid processes tend to be much smaller in females than in males.

There is evidence to suggest that if your mastoid processes point downwards, your earlobe will not be well defined (known as an "attached" earlobe). If the mastoid processes point forwards, you will probably have a well-defined, "free" lobe to your ear.

Determining how old a person was when they died is difficult from the bones of the neurocranium alone, unless the subject is a young child. Once adulthood is reached, the joints, or sutures, between each of the vault bones start to fuse, and they become really no more than a general guide to whether a person is young or old.

Sometimes, in the sutures between the different bones of the vault, accessory bones called Wormian bones may be found, which are a marker for some conditions, such as Down's syndrome and rickets. These are more common in some ancestral groups than others. For example, skulls of Asian Indian origin often contain an abundance of these extra little bony islands, while a very large, single extra bone at the back of the skull is often referred to as an "Inca" bone because of its prevalence among Peruvian mummies. This genetic predisposition to accessory bones in the skull sutures may provide valuable information in relation to ethnic origin.

There might also be small pits on the inside of the skull cap, running in lines parallel to the keel from front to back. These are caused by arachnoid granulations, protrusions of the membranes that cover the brain, passing upwards to drain into the long vein or sinus that runs from the front to the back, inside the skull cap. The arachnoid granulations, which look like little cauliflower florets, drain cerebrospinal fluid, the fluid that bathes the brain, from the spaces between the coverings of the brain up into the central venous sinus known as the superior sagittal sinus, allowing it to be recycled into the venous system.

With age, these depressions, known as granular foveolae, can become carved into the inner surface of the bone. If we see these, we may suspect that we are looking at the skull of an older individual. There was a fad at one time for trying to tell how old a person was by counting the pits, a bit like working out the age of a tree by counting the rings on its trunk, but foveolar counting is fantastical nonsense, even if it makes a good fable.

It is also possible to predict some forms of deafness from the skull. The ear has three distinct sections to it, all of which are formed through different processes. The external ear includes the pinna (the outer flappy bit on the side of our head) and an opening into a canal leading to the eardrum, or tympanic membrane, which lies inside the temporal bone. If the external opening to the skull is absent, no sound waves can reach the eardrum and the person will be deaf.

The middle ear, deep within the temporal bone, extends from the eardrum to the wall of the inner ear. Across this space, three small bones (ossicles) work together as a mechanism to transmit the vibrations from the eardrum to the inner ear. If the tiny joints between each of the three bones (malleus, incus and stapes—hammer, anvil and stirrup) are not functioning, again, the individual will be deaf. Fusion of the foot of the stirrup to the wall of the inner ear is another indicator of deafness. There are of course many other reasons why someone may be deaf, but this is some of the anatomical evidence we can read in the skull.

Deafness as a result of inner ear malformation (within the petrous part of the temporal bone) is trickier to identify and requires the anthropologist to be prepared to literally drill down into the very dense bone that developed around the embryonic otic capsule, the precursor of the inner ear. This is a fascinating little area of bone that is already formed to adult size at birth and, it is believed, does not remodel thereafter. The otic capsule is a gem for stable isotope analysis—analysis of the levels of elemental isotopes such as oxygen, nitrogen and phosphorus that can produce an elemental signature in our tissues. As this tiny bone is laid down from the building blocks of the mother's diet when she was pregnant, it can give scientists information about the food she was eating and the source of the water she was drinking at the time when her baby's inner ear was forming, which may in turn point them to whereabouts in the world she was living.

If a solitary skull turns up unexpectedly, regardless of how blindingly obvious it may be that it is human, the police must have this confirmed by a qualified expert before they can decide how to proceed. We were once sent a photograph of a skull that had been found by police on some wasteland. It was a very good copy, but it was apparent from the teeth that it was a cast. The fact that it turned up in November, shortly after Hallowe'en, may have been a clue to what it was doing there.

However, it is not unusual for isolated heads or skulls to be dredged up by fishing boats. When this happens it presents the skipper with a difficult decision, because if human remains are found in a catch, the whole catch has to be disposed of. So there are serious implications for their livelihood. For this reason, I am sure many finds must go unreported.

When a skull (minus its mandible) was noticed sitting in plain sight on a harbour wall on the west coast of Scotland, it was clear that a compromise had been found by one captain. It had obviously been brought up from the sea—there were barnacles attached to the surface—and placed there deliberately so that someone would report it to the authorities. The skull was photographed by a police officer, who sent us the picture requesting confirmation that it was human, which, of course, it was.

We were then asked to date the skull (in terms of estimating how long the individual had been dead), to define any distinguishing characteristics and to take a sample of bone for DNA analysis. That the skull belonged to a male was evident from the well-developed ridging above the eyes, the size of the mastoid processes and the prominent external occipital protuberance at the back. We believed he had been in his late teens to early twenties as his teeth showed limited wear. There was no dental work. The sutures had not started to close and, at the base of the skull, there was a gap still visible between his sphenoid and occipital bones. This gap is called the spheno-occipital synchondrosis (one of my favourite anatomical names) and it closes by around eighteen years of age in males.

The labs were unable to obtain a DNA profile from the bone. All things considered, our suspicion was that this was not a recent death. We sent a section of bone for $C^{14}$ radiocarbon dating and it was returned with an estimation that this man had been dead for six to eight hundred years. Whoever he was, bless him, he was not of forensic relevance. It was likely that coastal erosion had uncovered an ancient grave and the bones had been washed out to sea, only to be returned to shore via a fishing net.

Skulls that come in with the tide or are recovered from a catch of fish are often represented by the neurocranium alone. The facial bones are more delicate and tend to be damaged by dredgers or as they bounce around on the sea bed. A skull cap may be all we have, but there is still a lot we can tell from a brain box.

2

# The Face

## *Viscerocranium*

*"The face is a picture of the mind with the eyes as its interpreter"*

Cicero

Statesman, 106–43BC

There are two parts of our bodies that we are generally quite happy to have on public display at all times: our hands and our faces, both of which we use to express ourselves and to communicate. But it is the face on which we focus, and that we converse with, and therefore the face by which the majority of us instantly recognize each other.

However, in cultures where the face may be habitually covered, or when, for whatever reason, we are accustomed to concentrating on a different part of the body, it is interesting that our means of identifying our fellow humans adapts accordingly. Recently an oncology nurse told me that she had spent so much time over the years trying to find the veins on the backs of patients' hands that she had come to recognize them by their hands and jewellery as much as by their faces.

Not long ago I was invited to a conference in Riyadh by the Saudi Arabia Society for Forensic Science. This was my first visit to that part of the world. I was told it wasn't necessary for me to wear a burka, niqab or gloves, but out of respect for the local customs I donned the abaya, the traditional black women's robe, and shayla, or scarf, thereby courteously covering both my body and my hair but leaving my face and hands visible.

I actually found it quite comfortable to be dressed in the same way as the other women—it was like being part of a sisterhood, almost—and being virtually inconspicuous to the men at the conference. One of the western attendees had chosen not to adopt the country's dress code and, even though she was perfectly modestly attired, she received some quite hateful and vicious comments from fellow male delegates in the corridors of the conference hotel. They would hiss at her that she was a disgrace and she should cover her hair.

This was probably the first time I was ever personally conscious of gender hierarchy on a cultural scale. I have been very fortunate in my career to have been largely oblivious of any sex or gender discrimination. I put this down to the fact that my parents never told me I was a girl. Yes, my father expected me to be able to bake a good rhubarb crumble, but he also expected me to be capable of helping him French polish a dining table and shoot, gut and skin rabbits.

In the worlds of the military and the police, often viewed as misogynistic, I can honestly say I have never been aware of having been treated differently because of my double X chromosomes. It may just be that I am sufficiently bombastic or heedless of it not to have noticed, or perhaps I have just been lucky. The only two occasions when I suspected that my involvement was merely a nod towards "EDI"—equality, diversity and inclusion—were in fact both in academic settings. I handled those in a manner that ensured the two male senior managers concerned never gave me any further trouble. It helps being an anatomist: you can legitimately use terminology that is normal in our line of work but quite discomfiting for others. In both meetings, when it became clear that a question was being asked of me simply because I was the only woman in the room, I inquired, very politely, whether they were interested in my response or simply in the presence of my uterus. Of course they were mightily embarrassed, and assured me it was my opinion they wanted to hear. But interestingly, neither of them ever asked me a question in that way again.

At the conference in Saudi Arabia, women were required to sit on one side of the lecture theatre and men on the other, with a very clear demarcation between the two. It was here that I noticed something quite remarkable as I observed the interpersonal relationships between the women who chose to wear the niqab, and who were thus completely obscured except for their eyes. When they entered the room, I was surprised to see that they were able to recognize their friends from a considerable distance away, even though they were sitting down, their faces were covered, and they were all dressed in the same black clothing with no distinctive jewellery on show. I commented on this to a male Saudi colleague of mine, who could not explain how they were able to identify each other so easily. He invited me to his home to meet his wife and ask her.

My colleague's wife confirmed that she, too, had no trouble recognizing her niqab-clad friends but, as is often the case with skills we develop in infancy and take for granted, she could not pinpoint how she did it. We could only do what all good scientists do when they encounter something they can't explain: investigate. My male friend and I got a group of Saudi female scientists together and began to design an in-country experiment that would analyse the ability of Saudi women to distinguish between friends and strangers dressed in the full niqab.

Their first challenge was to assemble a large enough sample size. Even though the research team was entirely female, the culture of distrust among the potential participants hampered progress. Despite the team's adherence to all the research ethics, and the reassurance that the images required would be destroyed at the end of the study and that no third party would be given access to them, many of the women approached were nervous about having their photographs taken for identification purposes.

Using eye-tracking software, we wanted to analyse what women were looking at as they encountered other women in full veil, some of them known to them and others not, as a means of establishing the cues they were seeking to capture. We know from existing research that we identify familiar uncovered faces by focusing on the inverted triangle that delineates the eyes, nose, mouth and chin. Our group, however, had only the eyes, the overall shape and size of a person and their gait on which to base an opinion. It seems that when the face is covered, it is not just the eyes that are important triggers for identification, but also the imperfect ways in which we sit, walk or gesture.

As the study is still ongoing we don't have a definitive answer as yet, but if we get to the bottom of it, understanding and learning how to use this skill could prove extremely useful to organizations such as the security services.

The face, or viscerocranium, the smaller of the two parts of the skull, consists of three regions: the upper region for the forehead and eyes, the middle for the nose and cheeks and the lower area for the mouth, teeth and chin. The viscerocranium is where the tissues associated with many of our senses are housed, including sight, hearing, taste and smell. As these are formed before we are born, there is a controlled amount of growth associated with their development. The eye sockets are already large at birth because, as discussed in Chapter 1, the eyes form as a direct outgrowth from the brain and so mature very early.

The different working parts of the middle and inner ear are virtually adult-sized by the time we are born and our sense of smell is very well developed, although the collecting chamber for odours and aromas, the nose, will keep growing, like the external bits of our ears, throughout our lives. That's why old men seem to have such large ears. But the biggest growth occurs around our mouth, as most (though not all) babies are born without teeth.

By and large we are all pretty good at recognizing the faces of people we know, but research shows that we are largely rubbish at recalling the face of a stranger we have met only fleetingly. I am the constant butt of my family's humour as I frequently fail to remember people I have met many times. The most infamous example was at an office-warming for our lawyer's firm, where I introduced myself to one of the partners, only to be told that he had been a dinner guest at our house.

But even that pales in comparison to my legendary faux pas following my return from my second mission to Iraq. With Aberdeen airport fogbound, my plane was diverted to Edinburgh and my husband decided to drive there to pick me up. As I strode purposefully across the concourse, two excited little blonde girls came running towards me shouting, "Mummy! Mummy!" which, thankfully, was enough of a clue for me to swiftly recognize them as my children. Their father, however, was nowhere to be seen. He was, in fact, standing behind me, hands on hips, shaking his head in disbelief because I had just walked right past him. It is relevant to the extent of my embarrassment that by this time, my husband and I had known each other for over twenty-five years. I hadn't recognized him because he was sporting a goatee beard he hadn't had when I'd last seen him, which I have to say rather suited him.

I spend conferences staring at people's chests (not clever) trying read their name badges and I am sure there are people who must consider me a dreadful snob in the mistaken belief that I have deliberately ignored them. Such ineptitude is not only embarrassing but can only be seen as a significant failing in someone whose career is based on the identification of the human, or what remains of them. What can I say? Names stick in my mind, not faces.

There is a select group of individuals, to which it goes without saying I will never belong, who have well above-average ability to remember and recognize faces, even if they have only encountered them once. Most of us can remember about 20 per cent of the people we meet, but these "super-recognizers" can recall in the region of 80 per cent. Such an innate skill is, not surprisingly, in high demand in the intelligence and security world and also in the commercial market for private clients ranging from casinos to football clubs. The day may come quite soon when this human talent is replaced by automated facial recognition technology but until then, super-recognizers have proved hugely valuable to the police in cases as diverse as gang violence and sexual assault. Recently super-recognition was used to help identify the men behind the poisoning in Salisbury of the former Russian military intelligence officer Sergei Skripal and his daughter Yulia. The classification of super-recognizers emerged from an entirely different field of research: a clinical psychology experiment which was studying the opposite end of the spectrum: prosopagnosia. This is a clinical condition, sometimes described as face blindness, where people have extreme difficulty identifying faces. It can be enormously debilitating. A parent may not be able to pick their child up from school because they cannot recognize their offspring. Some sufferers cannot even recognize their own face on being shown a photograph of themselves. Prosopagnosia is an inherited condition; it can also be acquired through stroke or traumatic brain injury. You can take a quiz online to see where you lie on the prosopagnosia–super-recognizer spectrum. Most of us will be somewhere in the middle, with the vast majority proving better at recognizing their husband than I am.

However good or otherwise we are at recognizing our fellow humans, we can be briefly wrong-footed by natural changes in their appearance caused by ageing, or weight gain or loss, or by deliberate superficial alterations. Genetics, of course, play a significant role in determining how we look throughout our lives, but most of us modify our appearance to some degree on a fairly regular basis. We might swap our glasses for contact lenses, put on make-up, grow a beard or moustache or transform our hair colour. But these temporary cosmetic adjustments do not fundamentally alter the underlying structure of our faces. As a general rule, few of us will change to such an extent that we cannot be recognized by those who have known us intimately in the past. However, when we start to modify the framework, by, for example, shaving off the point of our chin, or acquiring cheek implants or veneers on our teeth, recognition can become more challenging. Such extreme forms of disguise have been integral to the plot of many a Hollywood movie.

Face transplants, once the domain of science fiction, are now a reality, if still a very rare procedure. Patients who have suffered severe disease, injuries or burns may be offered skin graft transplantation using tissue from a donor (including muscle, skin, blood vessels, nerves and, in some cases, even bone). In this surgery two fundamental alterations collide, with the creation of a new scaffold supporting somebody else's face producing something of a chimera. The operation neither restores the individual to their former appearance, nor bestows the appearance of the donor on the recipient. The result is a mix of the two, with the surgical process itself contributing significant additional alterations.

Such cutting-edge surgery is considered only when all other avenues have been exhausted. It carries a significant risk of rejection, which means that the patient must remain on immunosuppressants for the rest of their life, and involves many ethical, psychological and physical issues that will affect not only the recipient but also the donor's family and friends.

Face transplants are still a very new field—the first successful partial transplant was performed in France as recently as 2005 and the first successful full face transplant five years later in Spain—and to my knowledge none of these patients have to date come to the attention of forensic anthropologists. But it can only be a matter of time. It is just one more example of how crucial it is for us to remain open to the myriad possibilities surrounding successful identification and to approach each case free of preconceptions.

A disfigured face is extremely debilitating and isolating in a society that sets so much store by how we look. Anaplastology, the branch of medicine concerned with prosthetics, has been addressing more localized facial disfigurements since it developed as a specialism in the aftermath of the First World War in response to the need to help injured servicemen reintegrate with society. Replacement noses were probably the earliest prosthetic, required to repair faces ravaged by either warfare or syphilis. Prosthetics were originally carved out of inert materials, including ivory, metal and wood, which were gradually replaced by more realistic plastic and then latex alternatives.

Today, the sophistication of artificial eyes, noses and ears is exceptional. Noses can be designed to closely replicate the damaged version (unless the patient takes the opportunity to go for a new shape) and eyes and ears are painstakingly constructed to mirror the patient's other eye or ear, so that their face remains relatively unaltered and symmetrical.

Recognizing a face is one skill; being able to describe it is another. We are all familiar with the facial composites produced by the police from witness descriptions to help with the identification of criminals.

The different areas of the face are considered individually and then brought together to build the final face: forehead, eyebrows, eyes, nose, cheeks, mouth and chin.

Originally, likenesses were drawn by artists. Identikit, the first trademarked system using templates of separate features, was introduced in the United States in 1959. Subsequent methods such as Photofits and e-fits, involving photographs and electronic software, may produce more polished-looking results but they still rely on each feature being selected singly from a database of possibilities and then overlaid to construct the final composite image.

Nobody would claim that this can create a perfect replica of the subject. If you put together a face from, say, Angelina Jolie's eyes, Stephen Fry's nose and Eartha Kitt's mouth, you are bound to come up with something of a dog's dinner. The aim is to produce an image that has sufficient resonance with the viewer to elicit some intelligence that can be followed up by investigators. It is accepted that the accuracy of a composite face may be less than 50 per cent, which might not be considered terribly encouraging, but we must remember that sometimes it might be all an investigation has to go on. There is a tendency for the human eye to focus on, and the brain to remember, the unusual or abnormal. This can cut both ways. If an anatomical anomaly is present, and correctly described, it can prove to be an enormous help in the identification process. However, if it is wrong, it can send an inquiry severely off track.

Our recognition skills are, of course, normally only called upon in the context of identifying fellow living human beings. When it comes to recognizing the dead, our perceptions can be very different. Those of us who have sat with a loved one through their dying and death, or who have paid our respects to their bodies prior to their funeral, will be aware of how, when the essence of a person, and the animation and expression of their face, has been lost, the outer shell in which it resided often does not look quite like the person we remember. It is usually very much smaller and somehow empty.

Those facing the horrendous task of trying to identify someone who has suffered a violent or catastrophic death, or an individual who has been dead for some time, are going to find recognizing the person they loved more challenging. In the wake of the 2002 Bali bombings, around half of the bodies were incorrectly identified by their families, who had walked up and down rows of bloating, decomposed and fragmented corpses looking for their missing relative.

It is no surprise that, in such traumatic circumstances, many got it wrong. Their distress, the daunting mortuary environment and their urgent psychological need either to find, or not to find, their loved ones would all have contributed to their confusion. It is difficult to suggest to a close relation who insists they are 100 per cent certain, either for or against an identification, that they might be mistaken. It is for this reason that Interpol standards for disaster victim identification (DVI) stipulate that bodies should not be returned to their families on the basis of facial recognition alone. For a body to be released with scientific certainty, one of the three major identifiers is required: DNA, fingerprints or dental information.

When the face of a deceased person becomes unrecognizable, through decomposition or damage, we can reconstruct it in an attempt to identify them. Reconstruction is a tool in our forensic arsenal to which we frequently turn when all other avenues have been explored, and it requires a special set of skills that combines both art and science. The basic premise of facial reconstruction is the close relationship between our appearance and the morphology of the underlying bone and its covering of muscle, fat and skin.

Facial reconstruction can be achieved either by producing clay models or through three-dimensional computer modelling. The Manchester method, accepted today as the gold standard, and which I believe to be the most rigorous, requires us to have a skull, or at the very least a good cast or three-dimensional scan of the skull. Wooden pegs are glued, either physically or virtually, to the skull to indicate the thickness of the soft tissue that covers the bones at all points. This will vary according to the sex, age and ethnic origin of the individual.

Next, each of the forty-three or so muscles is added, one by one, layer by layer, to build up the underlying soft tissue scaffold as accurately as possible. The parotid glands, the major saliva glands, are also included at the side of the face, as are the buccal pads of fat in the cheeks. Then the skin is laid over everything and moulded into the contours of the face, much as you would lay a sheet of icing across the surface of a cake.

How the more cosmetic elements of the reconstruction are approached will depend on its purpose. Sometimes we do facial reconstructions for display, for example, with archaeological remains for a museum exhibit. With those models, the artist can be given reasonable leeway to add skin tone, eye colour, hair colour and style, facial hair and so on.

If it is to be issued to the press in the hope that it will help to identify a body, a greyscale illustration may be produced. Forensically, we may not be sure of skin colouration, and we do not want to guess at hair or eye colour as this could unduly influence the viewer to include or exclude a possible candidate.

The current research on DNA phenotyping may before long consign this uncertainty to the past. It is believed that we can now identify natural hair colour or eye colour from DNA. Other more complex features, such as eye shape, nose length or mouth width, may also have a genetic predisposition. They are more difficult to interpret, but it may well be possible, some time in the future, to partially reconstruct a face that looks like the living person from DNA alone.

Often, a simpler depiction may be sufficient to create a likeness that renders a disfigured or decomposing face acceptable for general circulation. This was the direction about to be taken by North Yorkshire police when I was approached to assist with the identification of a young female who had been found in the most unusual circumstances.

A couple of young lads had been out driving in the countryside when they spotted a silver suitcase dumped in a ditch at the side of a remote lane. Naturally, they stopped to have a good look. It was very heavy, and when they noticed it was starting to ooze a pungent brown liquid, they very wisely decided not to open it and to phone the local constabulary instead.

The suitcase was bagged and tagged and taken to the mortuary unopened because the police had a sneaking suspicion as to what might be inside. Their fears were well founded. At the mortuary, the police and the pathologist unfastened the case and discovered the body of a virtually naked young woman, curled up in a fetal position with her hips and knees bent to squeeze them into the tight space. Her face and head were wrapped in plastic tape. Her visible facial features indicated that she was of Asian origin.

Her DNA and fingerprints were run through the various databases but there was no match, and nothing seemed to tally with anyone listed on the UK missing persons register. Decomposition was not overly advanced, and the pathologist had determined that she had been dead only a matter of weeks. The cause of death was most likely asphyxiation.

The stage at which a forensic anthropologist enters the picture is often once an initial postmortem examination has been completed, the ensuing police investigations have not resulted in any solid new leads and progress is starting to falter. At this point we may be asked to perform a second PM to establish whether there is more information to be extracted from the body, which is what happened in this case.

The first examination is usually a hive of activity but by the time you get to the second PM, there is less bustle. I prefer it that way: the atmosphere is calmer and there is less pressure on you to perform. You might have a police photographer turn up, or you might not. The pathologist probably won't do more than pop in to say hello. It is usually just you and the mortuary technician. As a result, we get to form quite close working relationships with the anatomical pathology technologists (APTs), or mort techs. One piece of advice we always give our students is that you will never go wrong if you turn up at the mortuary bearing gifts. Biscuits are good (I bring biscuits with me wherever I go), chocolates are better, but jam doughnuts open all doors and melt the iciest of hearts. Trust me, you always want the APT on your side, and they never forget a kindness.

The disruption left by a first PM can take a bit of getting used to. The scalp will have been peeled back to gain access to the skull and the skull cap sawn off to excise the brain. The cavity is then usually filled with cotton wool and the scalp drawn back into place and stitched. The torso will show a T- or Y-shaped sutured incision that runs horizontally across the collar bones and vertically down to the pubic region.

When this is unstitched, there is generally a plastic bag inside the body cavity containing the brain and viscera, removed previously for examination or sampling for further laboratory testing. There is little reason for the anthropologist to open the bag of viscera: our interest lies in the external form and internal skeleton. It is quite common to find the back and the upper and lower limbs still intact, unless there is a trauma or pathology that has attracted interest and attention to those areas of the body.

Radiographs or even CT scans of the whole body may have been taken before the first PM and these, along with photographs from both the mortuary and the scene, give us as complete a background picture to the second examination as we could hope for.

The body may have been stored in a freezer, in which case it is usually taken out the day before the forensic anthropologist's postmortem to allow it to defrost. Mortuaries are not known for being cosy at the best of times, and working on a cold, wet, semi-defrosted body leaves your hands stinging with pain. This is where the doughnuts come into their own. The favour will be reciprocated with a hot cup of tea when you take a break and it is the best mug of warmth in the world.

What the police were keen to know from my PM on the young woman in the suitcase was how old she was and her ethnic origin. From the X-rays and my examination, I was able to establish that she was around twenty to twenty-five years of age when she died. Among other parts of the skeleton that provided me with this information were some small areas of bone around the edges of the breastbone (which we'll look at in detail in Chapter 4), and developmental changes we could see in her pelvis and skull.

I believed, from my assessment of her face and skull, that her ethnic origin was likely to be in the region of Vietnam, Korea, Taiwan, Japan or China. I didn't think she had the facial characteristics that would take her south into Malaysia or Indonesia. This deduction was based on the shape of her face, nose, eyes and teeth and on her hair type and colour. The suitcase was later established to have been manufactured in either South Korea or Lebanon.

However, none of this led us any further in terms of potential matches with missing persons and neither DNA nor fingerprint evidence was moving the case forward, either. We advised that the police should issue a black INTERPOL notice, the official international notification that a body has been found and not identified. The police had previously brought in a forensic artist, someone trained to produce a likeness of the human face that is accessible and palatable to the public—even when the real thing is showing signs of discolouration, decomposition and bloating, as this young lady's was—with the intention of releasing an image to the media.

Unfortunately, on this occasion, the melding of realistic portraiture and artistic interpretation had not produced a harmonious result.

While the talent of the artist was not in question, the upshot was a faithful reproduction of the face as she actually saw it, minus the decomposition, obviously. Bear in mind that the dead woman's face had been bound tightly by plastic tape and decomposition gases had caused bloating. The expanding face had therefore been constrained by the tension in the plastic. So although the final image was technically accurate, the impression it created was very odd. The victim's lips were ballooned in the midline and the crenulation of the top lip, where it had been pushed against the teeth, gave it a scalloped appearance. It didn't look like any mouth I had ever seen. My strong advice was that this drawing should not be released.

A more experienced artist would have made allowances for facial alteration due to decomposition. As it was, all people would see would be the postmortem consequences of the way the victim's body had been treated and it was highly unlikely that this image would lead to her being identified. Indeed, it might well prove more of a hindrance than a help. Fortunately, the police agreed.

Thankfully, as it turned out, the image was not crucial to giving her back her name: INTERPOL were able to confirm that they held a yellow notice, the notification issued for a missing person, relating to a twenty-one-year-old student from South Korea whose disappearance had been reported by her university in France. A communication with the South Korean embassy, a transfer of fingerprints from her ID card and her identity was swiftly confirmed.

Jin Hyo Jung had been visiting the UK as a tourist and had rented a room in a London flat owned by a Korean man. In the flat the police found a roll of plastic "Gilbert and George" gift tape, belonging to his girlfriend, which matched the tape wound around the face of the young student. Probably only 850 rolls of this tape had been sold by Tate Gallery outlets in the UK, and this one had his blood on it. Jin Hyo Jung's blood was also found in the flat and in the landlord's car, and her bank account had been emptied.

Often it is only during a trial that the full story of a crime emerges, and so it was in this case, when the landlord, Kim Kyu Soo, appeared at the Old Bailey. I was not required to give evidence on this occasion as her identity had been confirmed. During the trial we learned that a few weeks after the discovery of the body, the Metropolitan Police had become aware of a second South Korean student who had gone missing. A joint investigation had duly been launched by North Yorkshire Police and the Met.

The second student was eventually discovered, bound and gagged with the same tape, boarded up in a wardrobe in another of Kim's properties. He was found guilty of the murder of both women, theft from their bank accounts and perverting the course of justice by concealing their bodies. He received two life sentences.

I have used this case as a cautionary tale in teaching forensic art students about the importance of interpretation and of understanding the impact of the circumstances of a death on a facial depiction. When I show them a photo of Jin Hyo Jung alive alongside the postmortem drawing of her face, the response of over 90 per cent of students is that they would not consider them to be a match. They would therefore have rejected the possibility that this may be one and the same person.

I cannot say why the artist took the approach she did. Perhaps, as I supposed at the time, it was a lack of experience. Perhaps she was focusing on accuracy. Whatever the reason, it could have been a lead wasted, and it is a salutary reminder that even when a body is relatively fresh, we cannot rely solely on a likeness or facial recognition to bring us the information we need.

When a facial depiction is combined with a reconstruction in the hands of someone with experience and skill, the result can be eerily accurate. In a case from 2013, which readers of *All That Remains* may recall, it was a computer-generated image of the face of a murder victim, based on a CT scan of the skull, that led to the identification of a missing woman.

The discovery of human remains was made in a woodland clearing on Corstorphine Hill, on the outskirts of Edinburgh, by a ski instructor out cycling who had stopped for a breather. When he looked down at his feet, he thought he saw a dirt-covered face staring up at him from the ground. He recounted how he recoiled and had to take a second look, thinking that perhaps all he had seen was a root system that looked like a face, but he had been right the first time. He had stumbled upon a shallow grave that concealed the beheaded and dismembered body of a woman.

Analysis of the remains established her age, sex and height, blunt-force trauma injuries and the constriction of her throat by means unknown. But a careless comment at the scene from a scientist with no specialist knowledge of anthropology initially sent the police on a wild-goose chase in their quest to identify her. The "non-anthro" expert suggested that the woman looked "eastern European," perhaps Lithuanian, her cosmetic dentistry "vaguely Hungarian" and that she might be a migrant. This was a lesson in the inadvisability of relying on the uncorroborated hunches of armchair experts venturing outside their area of expertise. We have learned over time that it is best to say nothing at all until you are sure you cannot be overheard.

The assistance was then sought of more suitably qualified scientists from my department at Dundee University, who were involved in the analysis of tool marks associated with the dismemberment, in examining the remains for further information that might help to establish the victim's identity and in the depiction of the face, undertaken by my colleague Professor Caroline Wilkinson.

Using the computer method, the muscles and soft tissue were layered one by one over a CT scan of the skull and the skin stretched over the anatomical framework. Using the age assigned to the woman by the team, and available hair as a guide to length and style, Caroline produced a striking representation for circulation in the media that felt incredibly real.

Some jewellery, including a Claddagh ring—the traditional Irish ring featuring hands clasping a crowned heart—found on the body suggested a possible Celtic origin, so the police were advised to ensure that the facial image was also distributed across Ireland. And it was indeed in Dublin, not Lithuania, that the victim's family saw the reconstructed face on the news. The likeness was uncanny, and they contacted the Scottish police immediately.

The woman had been in Edinburgh visiting her son who, after DNA confirmed her identity, was arrested for her murder. The charge was reduced to culpable homicide on the basis of diminished responsibility. He was found guilty and given a nine-year custodial sentence for homicide and for the dismemberment and concealment of his mother's remains. The judge and three psychiatrists did not uphold his psychiatric plea for leniency on the grounds that he suspected his mother of being a reptile and wanted to look inside her to see if she was impersonating a human. As to why he cut her head and limbs off, dug a hole and buried her, he offered no explanation. Doubtless his reasons were more banal. Having mutilated her body, he decided to use a suitcase to carry it to the disposal site. Most murderers find it easier to do this if they separate the body into smaller parts.

The intricate manner in which the fourteen separate bones of the adult viscerocranium develop, grow and respond to our lifestyles is what creates the character in our faces. It is the skill of the expert in replicating the strong relationship between the underlying skull and the overlying face that makes the facial reconstructions so reliable.

Sometimes, all forensic experts have to work with is a skull and a possible lead to a missing person. In these circumstances, they may attempt a superimposition. This involves overlaying a photograph of the head of a missing person on to a photograph of the skull, taken in the same anatomical position. If the anatomical points (orbital margins, chin shape, cheek bone position and so on) can be aligned, it is possible to determine whether the skull "fits" with the face.

The first forensic application of superimposition is still viewed as the classic example. It helped to secure the conviction of Buck Ruxton, a physician who killed two women and was hanged in 1935 for the murder of one of them, his common-law wife. This investigation, which was notable for the use of various innovative forensic science techniques, is discussed in greater detail in the final chapter. It remains best known for the groundbreaking work of the pathologist, John Glaister, and anatomist James Brash in reconstructing the mutilated and decomposed bodies of the two women. The most famous image from the case is the superimposition of a photograph of one of the skulls on to the face of Isabella Ruxton. The incongruity of a composite combining a smiling face and a skull, crowned proudly by a diamond tiara, is unforgettably affecting.

Superimposition had a greater following in the past than perhaps it does today, simply because in the twenty-first century, scientific advances have opened up so many other avenues of investigation. But there are still times when we elect to go back to the methods pioneered by Glaister and Brash eighty-five years ago.

One of those arose in the mid-1990s, when we were assisting with a case that remains notorious in Italy. I was working as a consultant forensic anthropologist with the University of Glasgow, having moved back to Scotland from London, when I was dispatched to visit the carabinieri in Verona, charged with transporting some "material" back to the UK for analysis.

This being Italy, I met the police not in a featureless room with rickety tables and scuffed chairs, but over coffee in a high-end Veronese café. It is not by accident that the carabinieri are seen by some as Europe's most stylish police force. The officers recounted how, in 1994, a man called Gianfranco Stevanin had picked up a sex worker in his car in the northern town of Vicenza and offered her extra money if she would go home with him and allow him to take pictures of her.

They drove back to his remote farmhouse in Terrazzo, in the countryside south-east of Verona, where he engaged in several hours of increasingly violent sexual games. When the prostitute refused to continue, Stevanin held a knife to her throat. She offered him all her life savings if he would let her go and he agreed to take her home to collect the money. As the car slowed down at a toll booth she managed to escape and approach a stationary police vehicle. Stevanin was arrested for sexual assault and extortion and sentenced to two years and six months in prison.

This proved to be merely the start of the story that would unfold around the man who became known as the "Monster of Terrazzo." When police searched his house, they uncovered several thousand pornographic images of other women, thought to be prostitutes, files of detailed notes on them and items belonging to at least two of the women, including Biljana Pavlovic, a prostitute who had been reported missing the previous year. Alarm bells began to ring even more loudly when it became clear from one of the photographs, which showed significant violence to an intimate area of one victim's body, that she must have been dead at the time it was taken.

This was now a murder inquiry, and when, in the summer of 1995, a farmer found a sack containing a mutilated female corpse on land near Stevanin's house, the investigation escalated and heavy digging equipment was brought in to search the farm thoroughly. The badly decomposed remains of four more women were discovered, some with bags over their heads and ropes around their necks. The most pressing question was: who were they? Sex work can be a transient trade, and a haphazard lifestyle often goes with the territory. Girls appear on a particular "patch" for a while and may then move on without warning. Few will notice a missing prostitute, and their fellow sex workers are reluctant to talk to the police for fear of inviting trouble.

The police were now faced with the difficult challenge of linking the bodies with the photographs, descriptions and trophies in Stevanin's possession. The postmortems conducted by the pathologists had determined the sex and age of the women, and it was assigning names to them that was now of primary importance and the reason why I had been sent to Italy. They believed they had strong evidence that one of the victims was Biljana Pavlovic, and that another was likely to be a missing woman called Blazenka Smoljo. Both were eastern European and to date they had been unable to track down any relatives of either of them to gather any further information or samples for comparison.

In the café, the officers laid out pornographic and crime-scene photos on the table in front of me between the cappuccino cups. Looking at these horrific images in the beautiful city of Juliet's balcony and glorious open-air opera, as its citizens greeted one another cheerily and chatted over coffee and cake, felt somewhat surreal. For once I would have preferred to have been in a dingy police office where I could have looked at these pictures freely without fear of offending. But the carabinieri, it seemed, were not troubled by the sensitivities I was used to observing. The bodies were very badly decomposed and the police wanted to establish whether skull-to-photo superimposition, which they had neither the equipment nor the experience to undertake in Italy at that time, was a realistic possibility.

If I thought that meeting was surreal, I'd seen nothing yet. The upshot was a decision that the heads of the two victims they believed to be Biljana Pavlovic and Blazenka Smoljo, for whom they had photographs for comparison, should be transported to Scotland for analysis and superimposition. The heads were isolated from the corpses and sealed in a two white plastic buckets. To further conceal their contents, each white bucket was placed in a carrier bag bearing the name of a well-known, high-quality Italian designer. The bags were unceremoniously handed over to me, together with two letters, one in English and one in Italian, explaining what I was carrying and that I had the authority to do so.

The first hurdle came at the airport check-in desk, where I was told that madam could only take one piece of hand luggage on board and that the other must go in the hold. I duly produced the Italian letter. The woman behind the counter turned a little grey and issued my boarding pass without further comment. Now for security. I couldn't put my carrier bags through the scanner—think of the shock for the poor person looking at the screen—so I called the security guard aside and showed him my Italian letter. He turned a similar shade of grey and shepherded me through a side gate, bypassing the scanner.

As I boarded the plane, the lovely English flight attendant told me again that I must put my luggage in the hold. I handed over the English letter and explained that I could not do this because I was responsible for ensuring continuity of evidence. At least she didn't turn grey, but she did become most officious. She moved me into the almost empty business-class cabin, which I thought was very nice of her, until I realized that this was merely to isolate me from the other passengers. Far from receiving special treatment, I was effectively quarantined for the entire flight. Not so much as the offer of a glass of water. Unquestionably I was being viewed as undesirable and possibly even contagious. There was no warm goodbye as I disembarked at Heathrow, although I think I might have heard a sigh of relief.

The next dilemma I faced was at UK customs: should I declare or not declare? Having been brought up to be a good Scottish Presbyterian girl, I opted to declare. As I approached, the bored guard with his feet up on the desk looked up at me over his glasses to inquire whether the contents of my two designer carrier bags were "for my own consumption." After he read my English letter, he spluttered and ushered me away as swiftly as possible. By this point, I had travelled all the way from Verona to Heathrow and nobody had scanned or inspected my unusual cargo. I can't imagine that would be allowed to happen today. I sincerely hope not.

Now I had to get to Scotland. I queued up at security for a second time, clutching my English letter. The official told me that he didn't want to put my bags through the scanner but he would need to have a look inside them. At last! Someone was going to check. But as he started to lift the buckets out of my bags, I realized that he was intending to open them right there on the table in the midst of my fellow passengers and their belongings. I had to stop him and warn him he couldn't do this in public. We would have to go somewhere private, and with air-conditioning. These were heads, not skulls, and they still retained a lot of decomposing tissue that was very wet and smelly, and possibly even a few maggots. Up to this point his face had remained its normal healthy colour but suddenly it was visibly green. He hurriedly consulted his supervisor and directed me towards the departure lounge without taking so much as a peek.

When the next flight attendant read my English letter, he let out a squeak, bless him, raised his hands in horror and sent me to the back of the plane, where once again I was ignored for the whole flight. If he'd had the wherewithal to surround me with razor wire and give me a bell to ring while shouting "Unclean!" my pariah status could not have been more obvious. People were being moved forward into spare seats at the front rather than be asked to sit anywhere near me.

In Glasgow we defleshed the skulls, photographed them from all angles and performed three-dimensional scans. The images were oriented to match the poses in the photographs supplied by the Italian police.

Both skulls were female and from individuals of a similar age, so we could not separate them on that basis. Biljana and Blazenka had both been around twenty-four when they disappeared. The first skull we analysed did not correspond anatomically with the photo of Biljana but it was a good fit for Blazenka, and vice versa. Quietly confident that the two skulls were a match, we sent our results to Verona. The carabinieri asked us to hold on to them until the trial. Some weeks later, they confirmed that they had been able to obtain familial DNA for both women which supported our findings and it had at last been possible to formally identify them.

With identity established, technically, there was no need for me to give evidence at Stevanin's trial, but the public prosecutor did not want to miss the opportunity to infuse the proceedings with a little theatre by introducing a foreign forensic scientist and a method that would capture the interest of the media. And the skulls had to be returned to Italy anyway. If I gave evidence the court, rather than the police, would be paying my travel expenses, so the carabinieri, too, were keen for me to attend.

The return trip was more straightforward as my cargo now consisted merely of cleaned skulls, dry bone that was perfectly acceptable for anyone to check if they wished to, though again, all the airport and airline staff preferred to take my word for it. I was taken to the public prosecutor's house on the shores of Lake Garda for dinner, which was lovely, though I was nervous about my impending ordeal in a foreign court. My evidence would need to be translated and there was no telling what sort of questions I might be asked. I dressed smartly for my appearance, even if my shoes were killing me, and took my seat in the courtroom, the picture of transfixed terror.

There are few people in my life who have genuinely made my skin crawl, but Gianfranco Stevanin was one of them. As I stood in the witness box I tried very hard to ignore him, but his penetrating stare was disturbing, almost hypnotic. I gave my translated evidence and retired to a seat in the courtroom where I was able to observe the rest of the proceedings, even if I could not understand most of what was being said. At the end of the day the prisoner was escorted from the court. As he approached my seat he deliberately slowed his pace and turned his head to stare at me, long and hard. His mouth curved upwards in a wintry smile that never reached his cold, stony eyes. I felt my blood chill.

I knew there had been death threats against journalists who had spoken out against him and I did feel uneasy. For several months after the trial, I was more than a little jumpy when confronted by anything unexpected. This was the one and only time in my career when I was genuinely concerned for my safety and that of my family.

Stevanin's defence was that he remembered nothing of the sexual encounters with his victims due to a previous brain injury sustained in a motorcycle accident. For dramatic purposes, he'd had his head shaved so that the huge arc of a scar could be seen across his scalp. The defence lawyers had unsuccessfully challenged a psychiatric report that declared him mentally capable of standing trial. In January 1998 he was sentenced to life for the murder of six women, including Biljana and Blazenka.

The case led to a national debate in Italy on the question of the criminal responsibility of perpetrators affected by mental illness and the capability or otherwise of such individuals to understand the consequences of their actions. Stevanin's legal team capitalized on this to launch various further legal challenges but his sentence was upheld and the Monster of Terrazzo remains in prison in Abruzzo, where I believe he recently expressed a wish to become a Franciscan monk. Whatever comes of that, the world is a safer place with him behind bars.

How do we use the bones of a face to tell us something about the person it belonged to? Starting at the top, in the first region we have paired orbits, the eye sockets, that are largely, but not completely, symmetrical, separated by the root or bridge of the nose. The purpose of the orbits is to enclose and protect the eyeball and the six muscles that move it, a lacrimal (or tear) sac, nerves, blood vessels and ligaments, all surrounded by periorbital fat, which acts as a shock-absorber should the eye take a direct hit.

There are seven separate bones that make up the floor, roof and walls of the orbit: sphenoid, frontal, zygomatic, ethmoid, lacrimal, maxilla and palatine. They are all relatively thin and quite fragile. A projectile directed upward into the orbit will readily pierce the thin roof and enter the lower surface of the frontal lobe of the brain.

In the adult female, the rim of the orbit is quite sharp, whereas in the male it is more rounded, a distinction that can be used to start to build a tentative identity for the sex of an individual. In the male, the area of bone above the orbit (beneath the eyebrow) can become prominent, developing supraorbital ridging or even a shelf-like protuberance known as a torus. This is quite marked in some early human skulls and is believed to be formed by the dissipation of forces associated with the larger muscle mass of the more robust jaw. There is evidence that our jaws have reduced in size as our food has become softer and more processed. The ridging above the eyes and the root of the nose becomes more obvious in the male after puberty, with the significant increase in muscle mass caused by hormonal influences. Females tend to have little or no brow ridging and to retain a paedomorphic appearance.

Between two and six years of age, the frontal bone, just above the eyebrows, pneumatizes and air cells form between the two layers of bone. These air sacs coalesce to create the frontal air sinuses, spaces lined by a respiratory epithelium which produce mucus that will ultimately drain into the nose. Why these air sinuses form is not fully understood but what we do know is that the shape of the air space within the frontal bone is probably unique to every individual. This can be of value in confirming identity when we have access to X-rays of this area of a person's head in life which we can compare with our radiographs of a body. Interestingly, people with some congenital conditions, such as Down's syndrome, never develop these air sinuses.

This area of the face is prime territory for body modifications or implants. We often see piercings around the eyebrow, where a cannula has been inserted underneath the brow and out through the top to allow jewellery such as barbells and studs to be worn. Piercings may be vertical, horizontal or both, creating a T-shape. An awareness of modifications to the face is important in forensic analysis as they may be very relevant to identification of a body. There is no cavity that we will not examine, because even when the soft tissue no longer survives, the jewellery might.

Implants can sometimes be made in the eye itself, just under the sclera, the white bit of the eyeball. This part of the eye can even be tattooed by means of an injection of ink beneath the conjunctiva (the mucous membrane covering the eye and lining the eyelids) and above the sclera. This allows the normally white part of the eye to be changed to any colour you fancy, but it is a risky modification that can carry significant complications.

The position of the orbits makes it highly unlikely that the human viscerocranium will ever be confused with that of another animal and very clearly flags the human as a predator. Predators tend to have forward-facing orbits as this enables the eyes to work in stereoscopic vision, conferring the ability to judge depth. This is vital if you are a hunter and need to calculate how far away your prey is and how fast you will need to move to catch it. Animals whose orbits are on the side of their head are more likely to be prey than predator. Their optical priority is peripheral vision, so that they can watch out for the hunters. As the old saying has it, "Eyes at the front, the animal hunts, eyes on the sides, the animal hides."

The nose, which forms the middle part of the face, is wedged between the orbits and the mouth. The cheeks also sit in this plane, to either side. The nose houses the upper part of the respiratory tract and warms and moistens the air drawn in through the nostrils. Cold, dry air is painful to inhale. Our noses also have an important job as the gatekeepers of the lungs, helping to prevent foreign objects from getting into the respiratory system by trapping them in the sticky mucus that covers the hairs, or vibrissae, within the nostrils. The snottery green product of successful air filtration is familiar to us all, and especially to the little children it so fascinates.

As air is drawn in through the nose it circulates over the turbinates, or conchal bones (hence the word "conk"), which are highly vascularized and act like the folded metal slats seen in wall-mounted radiators. Having this massive blood supply within a structure that sits proud of our faces seems like something of a design fault, given that the nose is likely both to bear the brunt of any direct hit to the face and to suffer substantial blood loss as a consequence. Skulls are frequently found with broken nasal bones and a markedly deviated septum, often the legacy of knocks received in contact sports such as rugby or boxing.

The nose is also there to trap smells and convey them to the brain so that we can recognize them. At the very top of the roof of the nose is a small square of specialized mucosa, about 3 cm$^2$ in size, known as the olfactory epithelium. Incoming odours are dissolved in the mucus and olfactory nerve cells transmit the information through the cribriform plate, a sieve-like piece of the ethmoid bone, into the cranial cavity. From here the signal travels, via the olfactory cranial nerves, to the part of the deep cortex of the temporal lobe of the brain where the message is received.

It is the connections of the olfactory cortex to ancient areas of the brain such as the amygdala and the hippocampus that make some smells particularly evocative. I only need to catch a whiff of wood polish or turpentine to be transported back to my childhood days helping my father in his workshop. An impaired sense of smell is now considered to be an early warning of neurodegenerative disease and a factor in identifying those at risk of developing dementia. It also came to be recognized as a symptom of COVID-19.

Forensic science is very interested in the kinds of things that we may choose to stick up our noses. For example, the habit of line-snorting cocaine can be detected in both the hard and soft tissues of the nose and palate. The ischaemic, or vasoconstrictive, properties of the drug leave their mark on the tissue and can eventually lead to necrosis and even ultimate collapse of the nose. Usually it is the cartilage of the nasal septum that is most obviously affected but the damage can extend to the palate, making it difficult for the person affected to drink normally without expelling the liquid out through their nose.

Nasal washings can therefore prove to be extremely important to forensic investigations. Flushing the nasal cavities and collecting the fluid, so that pollen, spores or other debris can be retrieved, may provide vital information about the environment in which the deceased took their last breath. When we can identify pollen from a particular plant within the nasal cavities, this may well indicate that the deceased was killed in a different place from where their body was found.

This procedure can be quite fiddly, but between us, a fellow scientist, Patricia Wiltshire, and I devised a method that has proved effective. Pat, a forensic palynologist (a specialist in pollen, spores and other palynomorphs), was discussing with me one day how difficult it was for her to obtain nasal washings during postmortems, because it involved flushing saline up the nose and attempting to catch it before it disappeared down the pharynx. She had found it no easier to try to flush in the other direction, from the pharynx upwards, and she was still looking for a solution.

I was reminded that we had recently talked about how, during embalming, the Egyptians would remove the brain with a hooked wire inserted through the nose, and this gave me an idea. I suggested that, once the pathologist had removed the brain during the PM, thereby cutting the olfactory nerves, she could perhaps use the filtration properties of the cribriform plate of the ethmoid bone to get the wash down the nose from the brain cavity above. Apparently, this worked a treat and so a new method was born. Magical things happen at thresholds where two different worlds collide and one can offer a solution to a problem that has been vexing the other.

The nose and cheeks may provide clues to ancestry. The shape of the zygomatic bones of the cheeks can point to eastern origins, while the nose often helps us to differentiate between the high-bridged characteristics of some ancestral groups and the broad internasal dimensions of others. It is fascinating to sit on a train or on the London Underground noting the massive diversity in the shapes of the various components that make up human physiognomy and imagining the underlying skulls. You do get some funny looks, though, as it is customary for people not to look at each other on the Tube.

Facial piercings are also very common in the nasal area, around the bridge of the nose, into the cartilages at the side or through the septum. We also see them, to a lesser extent, around the cheeks, where dimple piercings, paired studs inserted into the high point of the cheek, seem to be a current trend.

It is in the lower part of the viscerocranium, the mouth and chin, that the greatest growth in the face can be seen as we mature and try to make space for all our grown-up teeth. The human is a diphyodont, which means that we have two sets of teeth: deciduous (baby or milk) teeth and permanent (adult) teeth. Of course, we are really triphyodonts, because, thanks to the skills of our dentists, our adult teeth are not necessarily permanent, at least not in the form they grew originally, and may be replaced by a third set of plastic or porcelain teeth.

It is, of course, a golden rule never to assume that any removable feature will necessarily end up with the person for whom it was originally intended. I remember, when I was working as a consultant in Glasgow in the 1990s, being present at the postmortem examination of the body of a tramp who had been found dead in the undergrowth of a local park. There were no suspicious circumstances surrounding his death: he was an elderly gentleman in poor health and, given that he had been discovered on a winter's morning after a cold night when the temperature had dropped well below zero, he had probably succumbed to hypothermia. But the police had no idea who he was, and it fell to us to come up with some clues to his identity so that any family could be notified.

He sported a full set of upper dentures (no lower ones) and, on the horseshoe area of the plate, we found a scratched reference number that could be used to try to locate the laboratory that had made the dentures for him, and, from there, perhaps lead us to the name of their owner.

However, it became clear as the investigation proceeded that the dentures worn by this gentleman had not been constructed for his mouth. We did indeed trace the laboratory and the man whose name they had on record was still very much alive. He had lost his dentures many years before, and it transpired that they may have gone through at least three identifiable owners before reaching the deceased. We thought it a testament to the famous Glaswegian iron constitution that the chap for whom the dentures had originally been made, betraying not the slightest hint of distaste that they had been found in the mouth of a dead man, asked if he could have them back as they were the "maist comfy wallies I ever owned."

Such potential mix-ups are perhaps not as rare as you might think. A nurse at my father's care home told me a story about an impish old lady who would go round at night collecting all the false teeth from the bedsides of sleeping residents and tumble them all into a sink (to "give them a good wash"), with the result that the next morning staff faced the time-consuming task of trying to match dentures to mouths, not always entirely successfully.

Teeth are the only part of the human skeletal structure that are visible to the naked eye and this makes them of significant value for identification purposes. They are also particularly useful in determining age. Tracing the development of the face of a child into adulthood is fascinating. Much of the growth is to do with making space for the maturing and emerging dentition, which is relatively painless and happens over a long period of time, but it can be clearly seen in photographic portraits of children taken once a year throughout their childhood. I did this with my own girls.

By two years of age the largely nondescript "baby" face is gone and toddlers will be recognizable as miniature versions of the people they will become. The twenty teeth that make up the full range of deciduous dentition have formed and erupted so their little faces have to be sufficiently mature to be able to accommodate them all. By six, the face has changed again, this time to accommodate the eruption of the first permanent molar at the back of each quadrant of the mouth. They will now have around twenty-four visible teeth, and there is a great deal more going on up in their gums that can't be seen.

There is a horrible phase between six and eight, during which the tooth fairy is disposing of the deciduous teeth and the permanent front teeth are erupting, when a child's mouth looks a bit like a robbed graveyard, with tombstones at all sorts of angles and stages of visibility. The face then transforms one more time, when the second molar erupts at around twelve, just ahead of puberty, before settling into its adult form around the age of fifteen.

The final teeth may be the ones that cause the most problems, especially if you already have a crowded palate. The wisdom teeth, so called because they do not appear until close to adulthood, by which time we are all supposed to have achieved some semblance of wisdom, must try to squeeze their way into a mouth in which all the other twenty-eight are already in place. Sometimes wisdom teeth do not form at all; sometimes they form but never erupt, and sometimes they choose to sprout at unacceptable angles and become the bully boys, pushing all our other teeth around. Their presence, then, can be variable, but when they are there, they offer the forensic anthropologist a clear indication of the maturity of the individual under examination.

All of the baby teeth erupt from the gum and fall out between the ages of six months and ten years. The permanent teeth start to push the baby ones out between six and seven and are all in the mouth by the time we are about fifteen. These well-defined stages of development make teeth incredibly important to estimation of age in the remains of a child.

The knowledge that teeth are shed in a relatively predictable pattern was put to good effect in 1833 as the government tried to establish fairer conditions for workers, specifically in the textile mills. The Factory Act stated that no child under the age of nine should be employed. Yet often their ages were a matter of guesswork, even for the children themselves, as in the UK there was no registration of births prior to 1837, and it did not become compulsory for almost another forty years. So age, and a child's fitness or otherwise to work, was established by looking at their dental development.

It was also acknowledged that no child younger than seven should be convicted for a crime as they could not yet be considered responsible for their own actions. The criterion used to determine age here was the eruption of the first permanent molar. If this had not yet happened, the child was deemed to be under seven and therefore below the age of criminal responsibility.

Forensic dentists, or odontologists, who specialize in the structure and diseases of teeth, are still employed today to help the courts determine the age of a child. Sometimes a minor who comes before the court, as either a victim or an offender, may not have documentation that confirms their age. Much of the world does not issue birth certificates and migrants and refugees who have fled for their lives do not always have their papers with them. In child slavery cases, any identity papers are often removed from the child, to render them utterly reliant on their "owners." To determine the age of these children, it is considered safer to look inside their mouths to assess their stage of dental development than it is to subject them to radiation with an X-ray, although nowadays, other options are available: their bones can be examined by imaging that uses non-ionizing radiation, including MRI scans.

Dentition can help forensic scientists to establish if, and for how long, a deceased newborn baby survived after birth. Birth is a fairly traumatic process, not only for Mum but for the baby, too. It disrupts the development of the teeth, resulting in a "neonatal line," a band which is microscopically visible in the enamel and dentine of the teeth formed up in the jaws at birth, and which is thought to be caused by the physiological changes that occur during the event. Since it appears only on teeth that are actively developing when a baby is born, it enables us to distinguish between prenatal and postnatal enamel formation. The approximate length of time the child lived can be calculated by measuring how much postnatal enamel has formed after the disruption that created the neonatal line. For forensic purposes, the presence of a neonatal line is accepted as an indicator of a live birth. When it is absent, it is likely either that the child was stillborn or that it died immediately after birth.

Our teeth change colour over time and can take on different hues when they are exposed to certain substances, which may offer pointers to identity. Children given antibiotics such as penicillin may be more likely to develop brown stains on their teeth. It has been suggested that this can also occur if the mother has taken antibiotics when pregnant. By contrast, an excess of ingested fluoride will result in white patches caused by fluorosis, a hypomineralization of the enamel.

Adult teeth may be dark in colour because of poor dental hygiene or staining by coffee, red wine, tobacco or other substances. Reddish black teeth might well be an indication of someone who chews betel nuts. This activity is prevalent in Asian culture and enjoyed by over 600 million people. In fact, betel nuts are the fourth most commonly consumed psychoactive substance after tobacco, alcohol and caffeinated drinks.

Today's dentistry attempts to counter these hazards by promoting the ideal of the same perfect Instagram smile for everybody: beautifully even teeth (or veneers) and bright white colouration. This is not very helpful to forensic dentistry, which relies on identifying the variations that occur naturally in our teeth as well as those resulting from dental intervention or restoration.

Following the Asian tsunami of 2004, forensic odontologists were able to confirm the identities of some of the deceased through matches to bleaching trays, fillings, root canals and bridges. The more dental work we have done, the easier it is for our teeth to be identified as ours, provided, of course, that our dental records are available for comparison. On the other hand, the more cosmetic interventions we opt for, such as having our teeth straightened by braces, the more uniform and less individualistic our smile becomes.

It is rather ironic that whereas during our lives we battle with tooth decay caused by what we eat and drink, after we die our teeth can prove to be extremely hardy. They can survive explosion and fire damage, shielded from such ravages by being encased within the mouth, and in many circumstances their longevity can outstrip that of our bones.

As a result, and because most people know a tooth when they see one, forensic anthropologists are often presented with isolated teeth. But recognizing a tooth is one thing; being able to determine whether or not it is human requires a level of understanding of the variability of dentition in a range of common animals. So more often than not, it is the molars of sheep, pigs, cows and horses that cross our desks. If it is a human tooth, is it one of the twenty we have as a child or the thirty-two we have as an adult? Is it an upper or lower tooth? Is it from the right or the left?

Teeth can tell us a lot about the life of the person or animal they belonged to from both a phylogenetic (or evolutionary) and an ontogenetic (individually developmental) perspective. We form the type of teeth we are going to need to manage our diet. Canines are essential equipment for carnivores but surplus to requirement for herbivores. Both need incisors and molars, but the molars are of a different type. The committed carnivore will have carnassial, or slicing, molars that act like scissors to snip off the pieces of meat it consumes, while the herbivore has grinding molars. As humans are omnivores, and eat a bit of everything, we have incisors to nibble and pinch, canines to pierce and molars to grind.

Sometimes the teeth that find their way to scientists are human, but turn out to be from a historic burial. The lack of modern dental treatment is an important indicator here but so, too, is the level of wear, which is not consistent with modern diets. High levels of carious lesions and decay tend to suggest a contemporary diet heavy in sugars whereas molars from archaeological remains are often eroded down through the enamel and into the dentine because of the amount of grit in the diets of bygone days.

Often it is that third, artificial, set of teeth that can prove the most fascinating, particularly in the shape of examples yielded by historical remains of the variety and ingenuity of the dentistry of the past. When, in 1991, I was part of a team excavating the crypt of St Barnabas church in West Kensington, London, the graves of three well-to-do women provided an insight into the everyday impact on their lives of their dental problems and of the efforts of their nineteenth-century dentists to solve them.

Sarah Frances Maxfield, the wife of Captain William Maxfield, who served in the navy for the East India Company, and in 1832 became the MP for Great Grimsby on the south bank of the Humber estuary in Lincolnshire, was buried in the crypt in 1842. She had been laid to rest next to her husband, who had died some five years previously. Other than that, all we know of Sarah is what we could surmise from the skeletal and dental remains within her lead coffin. She was evidently a lady of comfortable means, able to afford not only a triple coffin (a multilayered affair of wood and lead typical of the period for the wealthy), but also some expensive dental work during her lifetime.

When we exhumed Sarah, our eyes were caught straight away by the unmistakable glint of gold. On further investigation, we found that her right upper central incisor had been sawn off, probably then cauterized with acid, and her own crown then fixed to a solid-gold dental bridge. As gold does not tarnish, it was still shining brightly through the brown, soupy decomposition deposits within the coffin, almost 150 years after her interment. The bridge, which remained in position within her mouth, linked back to her right upper first molar, which was held in place by a ring, again made of gold.

Unfortunately, this tooth showed significant decay and the kind of extensive bone loss associated with chronic pus production, which was probably still active when she died. The only thing holding her molar in place was her dental bridge. The pain she must have suffered when trying to eat and the stench of decay can only be imagined.

Harriet Goodricke, who was sixty-four years old when she died in 1832, was also buried in a pricey triple coffin, but she had spent less than Sarah on her dental restorations. Harriet possessed a full upper-plate horseshoe denture, which had fallen out of her mouth by the time we examined her body. This was not surprising as there was nothing holding it in place. When the denture had been designed for Harriet, she must still have had a single tooth remaining in her upper jaw as the denture had a big hole in it on the right-hand side, corresponding to the position of her first molar, and the denture would have been constructed to fit over this tooth.

However, Harriet had subsequently lost the tooth and so there was nothing left to which her denture could be secured. It would therefore have served no practical purpose. It was a touching testament to the care of the person who had laid her out that the denture had been buried with her to preserve her dignity, and perhaps her pride in her appearance, even in death.

It must be said, though, that the denture would not have been overly convincing. It did not consist of separate artificial teeth but was carved out of a single piece of ivory (we could not determine the animal this came from: perhaps an elephant, but hippopotamus and walrus ivory was also commonly used in the nineteenth century), with the positions of the teeth delineated rather crudely by vertical lines, giving only a vague definition and approximation of the appearance of teeth. Dentures like this, which were fairly typical of the time, were often carved by clock- or watchmakers rather than anyone with a dental or medical background, and sometimes their anatomical accuracy left something to be desired.

Having been in the coffin for over 150 years, this bone denture had taken on the brown colouration of the viscous liquid in which it had resided (decomposition fluids that had mixed with the wood of the interior coffin to form a weak humic acid). Harriet's beloved accessory was therefore stained a very dark brown when we recovered it from her coffin and I am sure she would not have been impressed.

The Rolls-Royce of dentures possessed by this trio of women has to be the set sported by Hannah Lenten. Hannah, who was forty-nine years old when she died in 1838, was evidently a woman of considerable wealth. She occupied an ornate lead coffin and boasted some phenomenally costly and inventive dental work. Since replacement teeth of the kind Harriet had, which were frequently fashioned from ivory, were often less than lifelike, there was a demand for dentures featuring real teeth among those for whom money was no object.

Dentists would place adverts in newspapers seeking to buy human teeth. Sometimes these were supplied by the resurrectionists, or body-snatchers, active at the time. Others were scavenged from the mouths of dead soldiers (preferably young) who had perished on the battle-fields. After the Napoleonic Wars, these became known as "Waterloo teeth." They were often fixed to an ivory dental plate but Hannah's Waterloo teeth were riveted on to a solid-gold horseshoe—the ultimate in Victorian bling. When you consider that in the early nineteenth century even an ivory plate containing real teeth would cost in excess of £100 (about £12,000 today), you can only imagine what that denture would have set her back.

These extravagant creations were the brainchild of Claudius Ash, a silversmith and goldsmith who turned his hand to making top-quality dentures for the very rich. He became Britain's foremost manufacturer of dental appliances and by the mid-nineteenth century he dominated the high-end European market.

Because the multirooted molars in the back of the mouth are more tricky to extract than the front, single-rooted teeth, fewer of these tended to be removed. While there was an aesthetic imperative to replace the front teeth if possible, people were not as concerned about the less noticeable gaps at the back, and when substitutes were used, they were often crafted out of ivory or animal bone.

Hannah Lenten, though, had had six of her molars removed, and was the proud owner of both upper and lower dentures. In order to ensure that these would not cause embarrassment by falling out, the upper set was wired to the lower set by means of a pair of coiled gold springs and secured with swivel pins, so that when she opened her mouth the upper denture was forced upwards against the palate.

In total, Hannah's dentures comprised six front, single-rooted

Waterloo teeth, riveted by gold clasps to the solid gold horseshoe upper plate. The six replacement molars (three on each side) were made of ivory and, again, secured by gold rivets. Although her lower denture was only partial and made of ivory, it included six more human teeth not originally hers.

It was poignant to note that, even in an age where tooth decay was not treatable or preventable and it was therefore far more common-place for people to lose teeth, they were still sensitive about how this affected their appearance. So much so that these well-heeled ladies were prepared to go to a great deal of expense and discomfort in their quest to maintain their smiles.

Sarah, Harriet and Hannah, having held on to their prized teeth for a century and a half after their deaths, were now being removed, along with the other bodies buried beneath St Barnabas church, to allow the imperilled crypt to be repaired and restored. Their remains were cremated and their ashes scattered on consecrated ground, but their dentures live on as a record of the dentistry of the past.

While it is, of course, standard practice to check for artificial teeth when examining human remains, we rarely come upon such elaborate examples these days. It is, though, amazing how many other foreign objects can be found in the mouths of the dead. Piercings of the lips, tongue, of the spaces between teeth or even the uvula (the little pendant of tissue dangling from the back of the soft palate) are not unusual. We might discover gems embedded in teeth. We are even starting to see RFID (radio frequency identification) trackers implanted in the mouth. The ingenuity of the human, it seems, knows no bounds, and the ways in which we choose to modify or embellish our already unique faces, the part of our body through which we communicate with the world around us, is limited only by our imagination.

Our chin is another feature that is unique to the human and there-fore really interesting in terms of its purpose, variation and growth. What is the chin for? Is it about mastication, mechanics or communi-cation, or is it merely an evolutionary blip? In newborn babies, the two halves of the mandible are separate and fuse only during the first year of life. The chin grows substantially in small children to accommodate the roots of the incisor teeth, slowing down at around four years of age. In males, it shows marked alteration following puberty.

Chins vary endlessly in shape: they can be cleft, they can be double, they can be pointed (in women and children), generally squarer in men. They can therefore be of considerable help in determining sex from the skeleton, and sometimes in identifying individuals. Although they are a prominent potential point of contact for a fist, the bone is robust, so it takes quite a swipe to fracture the chin. But we do see such injuries.

The separate structures of the bones of the human face all play an important role in identification but it is only when they are brought together in perfect synchronicity that the composite becomes so much more than the sum of its parts.

# PART II

# THE BODY
## Postcranial Axial Bones

3

# The Spine
## *Vertebral Column*

*"You are only as old as your spine"*

Joseph Pilates

Physical Trainer, 1883–1967

I once wrote this very pompous academic opening to a chapter on the vertebral column: "The metameric segmentation of the central axis of the skeleton is the primitive phylogenetic phenomenon from which the subphylum vertebrata derives its name."

My colleague and friend Louise Scheuer told me I had incomprehensible verbal diarrhoea—I love her.

What I meant was that the human is formed around a central axis (skull and spine) and made up not of single bones but from many different segmental pieces that fit together, a bit like irregular-shaped toddler's building blocks. The fact that we have a spine, or vertebral column, is one of the features that defines us, as it is the basis for the animal classification of "vertebrates." If you don't have a spine then you are an invertebrate and probably not reading this book as you are likely to be an insect, spider, snail, crab, jellyfish, worm or some equally spineless character.

Like the word "spine," the name vertebral column has its origins in Latin (from *verto*, meaning to turn). Having a mobile spine allows us to be able to twist our bodies into many amazing shapes. As we grow older, though, the flexibility in the spine decreases and the movement we enjoyed in our youth becomes a distant memory. These changes can be seen in each of the bones of the spine individually and equally when we seriate, or stack, them into their correct anatomical position.

As we age, bony growths called osteophytes start to develop around the edges of each of the bones of our spine, which limits our movement and causes pain. Sometimes they can get so large that they fuse one vertebra to another, thereby permanently restricting our flexibility. The presence of osteophytes is useful for determining the possible age of an individual as they are rare in the young. They are one manifestation of osteoarthritis and can in occur in any and all of our twenty-four pre-sacral vertebrae.

We normally have around thirty-three vertebrae altogether: seven in our neck (cervical), twelve in the chest (thoracic), five in the small of the back (lumbar), five that fuse together in late childhood in the buttock region (sacrum) and around four that consolidate to form our rudimentary tail (coccyx).

When presented with a single vertebra, having established that it is human, the forensic anthropologist must decide which of the five regions it comes from, and which of the individual thirty-three adult bones it is most likely to be (there are almost three times as many bones in the spine of a newborn baby, before they start to fuse together). The answer may be pertinent to identity or hold other information that can assist the investigative authorities with the manner or cause of death. Sometimes all three.

A death by stabbing may well leave its marks on the vertebrae, but if all thirty-three become separated, perhaps when remains have been scavenged by animals or dispersed in water, we must be able to identify every bone in isolation and to list all those that are still missing and need to be found. Identifying individual vertebrae is, not surprisingly, a frequent topic for examination questions when students are learning their trade. They are usually asked to estimate the position of a single bone in the vertebral column and if their answer is more than one vertebra out in either direction, it will be marked as incorrect. We are tough.

Sometimes a single bone can be a mine of information about

who someone was—or wasn't. This certainly proved to be the case when my Dundee university team were challenged to verify whether remains interred in Wardlaw mausoleum, just outside Inverness, were truly those of Simon Fraser (1667–1747), the most notorious chief of the clan Lovat. The mausoleum, in the quiet hamlet of Kirkhill, was built in 1634 for the Lovat family and used by them as a burial vault until the early nineteenth century. Although permission for Fraser, the 11th Lord Lovat, to be laid to rest here was refused by the government of the day, legend had it that his body had been brought secretly to the vault from London.

Throughout his life, Simon Fraser, a wily rogue known as the Old Fox, opportunistically changed his allegiances to suit his own agenda. Having started out, ostensibly, as a supporter of the English Crown, he later defected to the Jacobite rebels loyal to Bonnie Prince Charlie. Inevitably, his duplicity eventually caught up with him and he was incarcerated in the Tower of London to await trial at Westminster Hall for high treason.

His was the biggest scalp to be taken by the Crown, and after six days of damning evidence, he was duly sentenced to the traitor's fate of death by hanging, drawing and quartering, which was subsequently commuted by the King to beheading. He thus acquired the dubious honour of becoming the last person in Britain to be executed for treason by this method. Thousands turned out on Tower Hill to watch the show. His demise was swift, if not without incident. One of the many overcrowded temporary spectator stands collapsed and nine people were killed. The Old Fox found this irony amusing and his reaction is said to have given rise to the expression "laughing your head off."

A drawing of the Old Fox was made by William Hogarth while Fraser was at the White Hart Inn, St Albans en route to his execution in London. It shows Fraser, an overweight, powerful and rather unpleasant-looking man, preparing to commit his thoughts to paper. His open diary and a quill pen lie waiting on a table beside him.

The government initially agreed that after his execution, Lovat's body could be returned to the family vault at Kirkhill (after his head had spent the requisite period on public display atop a spike as a deterrent to others). However, they changed their minds and decided that the Old Fox should be interred, along with two other Jacobite peers, the Earl of Kilmarnock and Lord Balmerino, in the Chapel of St Peter ad Vincula in the Inner Ward of the tower. But a story persisted that his body had in fact been smuggled out of London and taken up north by ship to Inverness and on to Kirkhill. Lovat was labelled by some as the last great Highlander, a true Scottish patriot, and his clansmen would not have wanted him to rest for ever on English soil. Today the mausoleum at Kirkhill is a tourist destination for devotees of the TV series *Outlander*, the fictitious time-travelling drama set in the spectacular local highland scenery around the period of the Jacobite rebellion, which has a cult following in the USA and Canada. I gather the Old Fox himself appears in one or two of the episodes.

There was some evidence to support the claim that Lovat's bones resided in the Kirkhill mausoleum. A marked space on the lid of a double-layer lead coffin in the crypt corresponded perfectly to an interesting bronze coffin plate that had become detached. The plate was inscribed with his name and his family coat of arms, in addition to an epigram in Latin denouncing the tyranny of neighbouring clans.

Two developments led to the involvement of my department in this ancient mystery. First, funds needed to be raised for the Wardlaw mausoleum to prevent the building from falling into disrepair, and secondly, the coffin lid was found to have been breached, which meant it would be only proper to exhume the remains within and transfer them to a sound vessel for safekeeping. As the 11th Lord Lovat was known to have been beheaded, we were naturally very interested in having the opportunity to examine the top of his spine, should it be present. But as things turned out, it was the bottom end that caused the early commotion.

Given the potentially significant findings from the analysis of the coffin contents, news of the exhumation caused a flurry of excitement. Dan Snow, the historian and television presenter, would be coming up to Kirkhill with a crew from History Hit TV to film the entire proceedings and the Royal Society of Edinburgh planned to hold a public event in Inverness at which the question of whether the Old Fox was buried in the Wardlaw mausoleum would be answered once and for all. The pressure was on.

I was accompanied to Kirkhill by my frequent partner in crime, my dear friend and colleague Professor Lucina Hackman. We trailed up there one cold day to view the site and start to plan our excavation. The mausoleum sits amid a beautiful old cemetery and is opened by means of an ancient key which apparently has its own starring role in *Outlander*. Every tourist wants to be photographed at the entrance holding that key, so we followed suit. Well, it would have been rude not to.

At the far end of a simple, rectangular room there is a trapdoor leading down to the crypt via a flight of steep stone steps. Laid out in this small, windowless, vaulted chamber (only in the centre is the ceiling high enough to allow you to stand upright) were six lead coffins, one child-sized, all belonging to the Fraser family, each bearing the name, age and date of death of the interred individual. All were intact except for the largest one—the one that had been breached—which was waiting for us in the far left corner of the crypt.

We donned masks, because it was clear from the dusting of white powder that the lead was oxidizing, and this can be a significant health hazard when you move the lead around and disperse the lead oxide particles into the air. We knelt down to look into the coffin through the narrow slits between the opened lid and the sides. We could see that there was a lot of wood present, probably the remnants of the inner wooden coffin. We could also see bone, so we backed away to agree a recovery strategy.

Since we wanted to try to get DNA from any bone present, we decided to go off and suit up fully, with double gloves to ensure that we did not contaminate anything. As to who did what job, there was no drawing of straws because Lucina always gets the short one (I am the elder). So Lucina would go down into the crypt to photograph and lift small sections of bone and coffin content, bit by bit. I would work upstairs in the mausoleum, where the remains would be relayed to me by a runner for further photography, recording, sampling and analysis. As we lifted the lid, it was apparent that the inner wooden coffin had collapsed, leaving some bones lying on the surface. The first to be pulled out was a sacrum, the large, triangular bone at the base of the spine. It was robust and in relatively good condition.

We could tell quite a lot just from this one bone. First, it was, most likely, that of a male. We can determine that from the shape and relative proportions. Simply from its size we could say that it belonged to quite a big male. He would have been elderly when he died, judging by the extent of the arthritis at various joints. From Hogarth's sketch and contemporary accounts, we knew that Fraser was tall (nearly six feet), and a man of considerable girth. He was around eighty years of age when he was executed and had suffered from gout and arthritis. So far, so good. The first bone out of the coffin seemed to confirm many of these descriptors.

In his excitement Dan was all set to declare that we had found the body of Lord Lovat. Much as we hated to rain on his parade, we had to remind him that we really needed to wait to see what else was in there before jumping to any conclusions. Dan decided to leave us to it and disappeared off to do some filming near Culloden battlefield.

The second piece of bone to appear was from the left knee region of an adult femur (thigh bone). This showed no sign of arthritis, which made us suspicious that it might not belong to the same person as the sacrum. It was the third find that confirmed beyond all doubt that what we had was a commingled burial. At the head end of the coffin, Lucina removed seventeen tooth crowns that had come from a child of about four years old. We had no idea why these might have been in the coffin, or of the whereabouts of the head to which they belonged. They may just have been a collection of baby teeth kept by a mother, which had to be put somewhere. Teeth can end up in some strange places. The tooth fairy let me hold on to all my children's first teeth and eventually they became part of a scientific experiment on estimating age from teeth.

We also uncovered the ribs and sternum (breastbone) of another small adult. They had been placed in an anatomically correct position under the base plate of the wooden part of the coffin, down at the foot end. Now, where was the rest of that body? We had no answer to that question. What we could say was that we were looking at the partial remains of four separate individuals: child, small adult, mature adult and elderly male.

But there was more. Lying along the bottom of the coffin, on top of the wooden base plate, in an articulated anatomical position, was a very poorly preserved skeleton—minus its skull. Lucina had come upstairs to give me the news of this find in a whisper, to avoid raising anybody's hopes too high. We also wanted to keep this information quiet until we had all our ducks in a row and could orchestrate a large-scale, public revelation of our conclusions. It was looking as if the clansmen had fooled the English in the end and had managed to bring their chief home to Kirkhill after all.

On his return, Dan expected to hear that we had identified the remains of an elderly male, as suggested by the sacrum we had examined. The name of this bone is an eighteenth-century abbreviation of its Latin name, *os sacrum*, or sacred bone. In English and German it was also known as the holy bone. Quite why it was considered holy is open to interpretation. One theory is an ancient belief that, being strong and resistant to decay, it will form the basis for corporeal resurrection on the Day of Judgement. Another is that it is a reference to its protection of the sacred organs of reproduction. Whatever the etymology, Dan had his hopes pinned on this particular sacrum providing the confirmation he wanted of the Old Fox's cunning plan for his body to be snatched from beneath the very noses of the Crown and transported home in glory.

When we broke it to him that there were at least five people in the coffin, he was aghast. He asked us how this could have happened. As the coffin had been breached, we believed that what we were probably looking at was graveyard tidy-up time. Pretty much whenever an animal or a human digs a hole in a cemetery, a bone will come to the surface. And when you find them, you have to do something with them. The obvious solution here was to slip them through the opened edge of a coffin in the crypt, where they would remain on sacred ground. It is the graveyard equivalent of sweeping dirt under the carpet. It is likely that the ribs and sternum were put in the coffin before the headless body, and that the other pieces had been dropped in after the lid had been breached.

We cleared the crypt of everyone who did not need to be there so that we could film our discussion of the headless individual we had found in the coffin. Everyone present was sworn to secrecy until the big night of the Royal Society of Edinburgh's lecture. All the tickets had been sold and Dan's TV company was live-streaming the event across the world to *Outlander* fans. Around four hundred people were in attendance and over half a million tuned in on TV that night or have seen the film since. It was the biggest audience the Royal Society of Edinburgh had ever had in its history of public engagement. The Old Fox could still draw the crowds.

Several journalists, reasoning that there was no way we would have gone to all this trouble if the Old Fox was not in the crypt, tried to get us to tip them off on what we had discovered, but we remained tight-lipped. On the night of the lecture, you could really feel the electricity in the room as we deliberately built the tension. Eric Lundberg, the custodian of the mausoleum, provided some background to the exhumation; Sarah Fraser, the celebrated historian and author, who married into the clan Fraser, gripped the audience with her talk on the importance of her family's ancestor in the time in which he lived. To sprinkle a bit of showbiz glitz, Dan Skyped into the event to tell everyone what this investigation meant to him and to show film clips of the progress of the exhumation.

Then Lucina got to her feet to set the scene in the crypt before I revealed what we knew of the headless person in the coffin. You could have heard a pin drop as I announced that if Lord Lovat was a twenty- to thirty-year-old woman, then we had indeed found him. The gasp around the room was audible. I genuinely don't think anybody expected this outcome. But that is the nature of science. It doesn't mould itself to accommodate the desires of humankind, it is there to convey truth.

We now had to explain our findings. The honest answer was that we did not, and still do not, know who this woman was. One suggestion is that a coffin, with its appropriately inscribed name-plate, was prepared for Lord Lovat but that ultimately the mission to spring his body from the Tower of London failed. Did the family simply remove the plate from the lid and use the coffin for somebody else? If they did, then they did not give this woman the courtesy of recording her name. How do we know she was a female? The shape of her own sacrum, and of her pelvis, left us in little doubt.

As Lucina and I had theorized, it was likely that the coffin had become a handy receptacle for various isolated finds elsewhere in the graveyard once some curious person had first opened it to have a look inside. Given the myth of the repatriation of Lord Lovat's body, it is also likely that in the intervening 250 years, others couldn't resist having a peep into the coffin—if the lead had become fragile and the soldered edges started to sag, it would have been even more of a temptation—and that this was responsible for the damage.

So where was the woman's head? There was no evidence that she had been beheaded, it was just that her skull was no longer there. Had one of these inquisitive people perhaps removed it? Did they look in the coffin and think, if this is the Old Fox, then that shouldn't be here, and take away the skull to keep the myth alive? Or did they believe it was in fact Simon Fraser's head and steal it as a trophy? We will never know.

Whoever the person in the coffin was, she deserved the dignity and decency of a secure reburial. My family has long known another branch of the Fraser clan, a renowned dynasty of funeral directors in Inverness. I phoned Bill and Martin and asked if they might be willing to donate a coffin for someone who was perhaps a Fraser, too, so that we could replace her remains in the crypt of the mausoleum. Of course they agreed, and with due solemnity the bones were reinterred and a service held. So the crypt now houses all the original lead coffins, plus a shiny new wooden one containing what is left of an unknown female, together with the bits of the four other people who have kept her company for generations. Knowing what we do of the Old Fox, I think if he was watching this story unfold, he would find it mightily amusing that over 250 years later he was still calling the shots. Probably laughing his head off.

The quest for the body of Lord Lovat may have begun and ended with the discovery of the sacrum of a well-built, elderly, arthritic male. However, it was not only the recognition of individual bones, but the ability to interpret information about the sex, age and other characteristics of the lives they had sustained that held the key to solving at least one element of the mystery and to figuring out how many bits of different people shared the coffin.

Our vertebral column, like that of all mammals who habitually walk on four limbs, was originally a horizontal structure. It was over 4 million years ago, according to the fossil record, that the ancestors of the modern human started to spend more time on two legs than four and the spine became vertical rather than horizontal. This was largely a bad idea, biomechanically speaking, as it placed tremendous compressive and tensile strains on the spine, with the result that most of the afflictions of our old age reflect a life of locomotion for which the axis of our bodies was simply not designed.

It should come as no surprise that when small children begin to become mobile, they start off on four legs, with their spines horizontal, the posture in which they are the most stable. When they start to become bipedal their movement is very tentative and wobbly until the muscles, bones and neurology get to grips with the ridiculous notion of such a small centre of gravity positioned precariously over two tiny feet. Perhaps it should come as no surprise, either, that when adults become neurologically impaired, perhaps after one too many libations, reverting to all fours proves a much safer, if less elegant means of locomotion, especially when going upstairs. Some babies, of course, go through an intermediate stage in which the ischial tuberosities— known to most of us as the sitting bones—offer a steadier means of locomotion and they become bottom-shufflers before finally dispensing with the stabilizers and taking to their feet.

The core purpose of the backbone in all vertebrates is to protect the incredibly delicate spinal cord and its coverings, which pass down the length of the body from the brain. At the top, this nervous tissue still consists of brain stem, officially becoming the spinal cord by the second vertebra in the neck. This remarkably thin, white cord, in its bony tunnel, carries all of our motor information, which instructs our muscles to work, while the sensory messages of touch, temperature and pain go in the opposite direction from the body up to the brain. The vertebral column is longer than the spinal cord, which stops short of the sacrum and coccyx in the lumbar region, around the first or second lumbar vertebrae.

This is why a lumbar puncture, or spinal tap, is carried out between two lower lumbar vertebrae (usually L3 and L4). This allows a frighteningly long needle to be inserted between the bones to get at the cerebrospinal fluid surrounding the nervous tissue without running the risk of hitting the spinal cord itself. Having experienced one of those myself while being tested for meningitis, I can attest that it is incredibly unpleasant, especially when the doctor, knowing that you are an anatomist, provides a running commentary listing every tissue the needle is penetrating. "Oh, there goes the posterior longitudinal ligament. Did you feel it pop?" is not something you need to hear.

When we decided to stand up on two legs, we asked our vertebral column to do some things it simply wasn't built for. As well as needing to balance the whole of our upper body on the lower limbs and our head on our neck, we required sites of attachment for the muscles controlling posture, with a sensitive nerve supply that could tweak them continuously to balance flexion with extension and keep us upright.

We are mostly unaware that we are constantly performing this delicate balancing act. It is a subconscious activity because, frankly, we can't be trusted to remember to keep doing it all day. When it is noticeable is when we fall asleep in an upright position. Humans can fall asleep standing up but they will quickly topple over if not supported when the muscles of postural balance become inactive. If you are in any doubt, watch as someone drops off to sleep on the sofa. That "nodding dog" impression they do when they jerk awake again is their subconscious nudging their conscious to tell it that the muscles of the neck have relaxed.

Like the bones around the brain, the vertebrae surrounding the spinal cord start to develop very early on in fetal life, within the seventh week, and, as might be expected, the bone formation starts at the top of the column, closest to the brain. By the time the baby is born, the spine will comprise nearly ninety different little bones that look like the jacks used in the game knucklebones (which is no coincidence, as the game was originally played with sheep vertebrae). The column grows so swiftly that by the age of four the separate pieces of bone will have fused together and consolidated into our thirty-three vertebrae, with the five sacral bones finally fusing together in a block in late childhood.

The fetal vertebral column curves in a C-shape with the concavity to the front. But something quite miraculous happens around two or three months after birth. The muscles in the baby's neck start to strengthen and it becomes able to support and balance the weight of its own massive head on the top of the spinal column. That spinal curve begins to reverse in the cervical vertebrae and becomes more convex towards the front of the neck. At around six to eight months, the muscles in the lower back develop and the baby is able to sit up unaided, balancing its whole body as it sits. This results in a further alteration to the shape of the column in the lumbar region as it, too, starts to become convex towards the front.

Before the candles have been blown out on baby's first birthday cake, the whole vertebral column has been transformed from the fetal C-shape into an S-shape, something seen only in bipedal animals. This shape is maintained by changes not to the bones, but to the pads of cartilage, the discs, that sit between them. As we start to age, the process reverses as the discs lose their elasticity and begin to collapse, and our vertebral column starts to revert to the C-shape of our formative years. We lose our ramrod-straight structure and become increasingly hunched and bent over. This shifts our centre of gravity, which using a walking stick helps us to stabilize.

The neck, or cervical, region, where the vertebrae help us to balance our head on our column, is very flexible. The shape of the vertebrae allows extensive rotation, so that we can turn our heads to look over our shoulders and nod it up and down.

The vertebrae that make up the chest region of the column provide sites of attachment for the ribs. These are the vertebrae most likely to show evidence of fracture in the elderly as a result of osteoporosis and those responsible for the rather cruelly named "dowager's hump." It is this area of the column that is most likely to fuse in old age as the bony bridges of osteoarthritis link adjacent bones and limit movement. These changes are most commonly seen for the first time in individuals in their fifth decade but they can happen much earlier.

The vertebrae of the upper thoracic region are not symmetrical. Each has a little flattened area where the aorta, the body's largest artery, lies next to the bone. In those who have died of an aortic aneurysm—a ballooning of the aorta which results in its walls becoming thinner and thinner until they eventually burst, as my Uncle Willie's did, very suddenly, at our Sunday lunch table—the traces of the aneurysm can be visible on the upper thoracic vertebrae even after the soft tissue is long gone.

The vertebrae in the small of the back are the largest of all the bones in the spine because they need to transfer all the weight of the body down to the sacrum and then to the lower limbs and on to the ground. Sometimes the last lumbar vertebra doesn't form properly and the different parts don't fuse as they should, resulting in spondylolysis. This condition turns into spondylolisthesis when the two parts are forcibly separated, which can happen while we are performing the simplest of actions, such as putting the cover on to a duvet (as my husband discovered to his cost). Sometimes we ask too much of the vertebral column and when it finally rebels, the consequences can be spectacularly disabling.

Around puberty, the five separate vertebrae of the sacrum will have formed a single bone. What happens to our coccyx, or tail bones, is very variable, in terms of both fusion and the size they become. The human does not, of course, have a free-hanging prehensile tail; instead the terminal bones of the coccyx are tucked under the natal cleft, the deep groove that runs between our buttocks. They are an important anchor for the attachment of ligaments and muscles. It is essential that, having decided to stand up on our two legs, we have a strong pelvic floor: this acts as a sort of anatomical hammock to prevent our guts from falling out of our bottoms.

Most of the time, all of this anatomical busy-ness around the vertebral column runs smoothly but—not surprisingly, given that the body is having to co-ordinate the development of ninety little pieces of bone—occasionally things don't go as anticipated. Some vertebrae may not form properly (resulting, for example, in butterfly vertebrae); others might fuse together when they shouldn't (diffuse idiopathic skeletal hyperostosis, or DISH), and some will stay apart when they should have fused, which is what causes spina bifida. Such anatomical variations, some of which go totally unnoticed by those who possess them, can help the forensic anthropologist to find some evidence as to who they might have been in life, especially when what we find can be corroborated by previous medical imaging.

The first two cervical vertebrae, at the top of the column, are of particular interest and anatomically very different from the rest. The first, C1, is almost a circle of bone. This is known as the atlas, after the Greek Titan who was condemned by Zeus to hold up the heavens on his shoulders for all eternity. In humans, the atlas has only to hold up the head, which is no mean feat in itself. The joint between C1 and the skull is highly specialized: it is this that enables us to nod our head.

The second cervical vertebra, C2, or the axis, is a very unusual-looking bone with a peg protruding from its upper surface. This peg, the odontoid process, fits inside the circular atlas. Ligaments wrapped around the waist of the peg allow it to rotate so that we can turn our head from side to side. Ingenious engineering.

Because these two vertebrae are so close to the skull, and to the brain inside, they have a very large volume of neurological tissue to protect, which is why the hole in the middle of each bone, the spinal canal, is very large. And this means that any damage or trauma this high up in the column can have fatal repercussions.

One of the most evocative labels that could ever be given to a fracture is one that involves the second cervical vertebra. It doesn't take a genius to figure out how the "hangman's fracture" got its name. Because the axis is fundamentally a ring, when it breaks, it must break in two places. If you doubt me, try to create a single break in a certain well-known mint with a hole in the middle. There will always be two pieces.

When somebody is hanged, the fractures of C2 tend to occur on either side of the odontoid process. They are caused by the sharp drop and the sudden, jolting halt of the rope. The peg ploughs backwards into the spinal cord (to be exact, the lower aspect of the brain stem) and the disruption to the neurological tissue causes death—virtually instantaneously, if you are lucky, and if the executioner does a good job.

Many hangmen took pride in a "good" hanging. It was a nineteenth-century British hangman, William Marwood, who, in 1872, designed the more precise "long drop," which took into account factors such as the condemned person's height and weight to calculate the optimum length of rope and drop to try to ensure the cleanest and most humane execution possible for each individual. Botched hangings were not only a cruel death, but distressing for the hangmen, and for those required to witness them.

The aim was to break the neck to cause instantaneous death but not decapitation. Yet, however well calculated, not all executions using this method were successful. Research has shown that fewer than 20 per cent of judicial hangings resulted in cervical fracture and of these, only around half would display the classic "hangman's fracture." Ironically, then, it is not actually "classic" at all but quite infrequently achieved.

This is why the sentencing would state that a condemned prisoner was to "hang by the neck until dead": the drop was sometimes not sufficient in itself to cause death and suffocation and vascular constriction with ensuing hypoxia were more likely—hence the "hangman's jig," where struggling could go on for several minutes before death occurred. To help bring about a swifter end, family or friends of the condemned prisoner might pay the hangman, or someone else, to pull on the feet. The placing of the knot on the noose could also be instrumental in bringing about a quick death. A submental knot, under the chin, helped to hyperextend the neck and so aid the fatal crushing of the brain stem.

Technically, hanging and strangulation are not necessarily synonymous, although hanging may result in death by a sub-category of strangulation if it is not instantaneous through neurological trauma. Strangulation, which is defined as asphyxia due to constriction of major blood vessels, inhibition of the vagus nerve or obstruction of the air passage in the neck, is a soft-tissue injury, and usually the domain of the pathologist rather than the anthropologist, as strangulation generally leaves no traces on the vertebrae.

The two other methods of strangulation are ligature strangulation, which can either be caused by an assailant or self-inflicted, and manual strangulation, using the hands (and sometimes other parts of the body) to fatally constrict the neck. Manual strangulation is highly unlikely to be self-inflicted for obvious reasons. The distinction between the three types lies in the cause of the external pressure on the neck. A constricting band exacerbated by the weight of the victim's body is the result of hanging. Ligature strangulation is defined by a constricting band tightened by a force other than body weight and manual strangulation is when the constriction is caused by the hands, forearms or any other part of the body.

There are also three types of hanging: free suspension of the body, incomplete suspension and hanging brought about by a fall from a height (most often due to judicial hanging). It is only in this last category that we are likely to see fractures in the upper cervical vertebrae that relate to the cause of death. Any other form of hanging may leave no mark on the bones.

All of these varieties of hanging and strangulation, except for manual strangulation, may be the result of suicide as well as homicide. A classic hangman's fracture was the outcome for William Bury, the last man to be hanged in Dundee in 1889. He had been found guilty of the murder of his wife, Ellen, but there were claims that her death may have been due to ligature self-strangulation and partial suspension.

Bury's murder conviction was not the extent of his notoriety. The timing of his arrival in Dundee from the East End of London, together with certain aspects of Ellen's demise, led to speculation in some quarters following his arrest that he may have been none other than Jack the Ripper, England's most infamous serial killer. However, with little evidence to support this theory, he would be some way down my list of likely candidates.

Bury and Ellen left Bow, not far from Whitechapel, Jack the Ripper's stamping ground, on 20 January 1889, a couple of months after the last of the five murders most strongly attributed to the Ripper. They sailed north on the SS *Cambria* to Dundee, where Bury had convinced Ellen that there was a job waiting for him in the jute mills. He lied. They rented a top-floor flat at 43 Union Street for eight days, until they ran out of money. Bury then put a roof over their heads—a dingy, unfurnished basement flat at 113 Prince's Street—by claiming that he wanted to view the property and simply failed to return the keys. Twelve days later he walked into the local police station and told them that if they went to the flat, they would find Ellen's body inside a wooden trunk. It was established that, by the time she was found, she had been dead for around five days.

Bury was known to be a drunk. He was also known to have been abusive and violent towards Ellen. It was suspected that he had married her because she had a little money from a small windfall, now long gone. What we will never know for certain is why he chose to come to Dundee and what happened in the basement of 113 Prince's Street on 5 February 1889.

Bury told the police that he and Ellen had been drinking the night before she died, although by all accounts, Ellen was not much of a drinker. He said that he went to bed and when he woke up the following morning she was dead, lying on the floor with a cord around her neck. It transpired that Bury had bought the cord the day before Ellen's death. It also emerged that on the same day he had spent some hours at the Sheriff Court, listening to proceedings from the public gallery, perhaps doing a bit of research on the legal system.

The Crown pathologists, Drs Templeman and Stalker, carried out a postmortem examination on Ellen's body and found several bruises and cut marks, one of them so deep that her intestines were protruding from her abdomen. They suggested that the edges of the wound were raised, an indication that she might have been alive when she was disembowelled. Around her neck were impressions left by the cord biting into her skin and the bones of her right leg had been broken in order to squeeze her into the wooden trunk where the police had discovered her mutilated body, packed around with clothing and books.

They also found a knife on the window ledge, the blade still encrusted with blood and hair that matched Ellen's. The picture seemed very clear: Bury had strangled his wife with the cord, and while she was still alive he had taken a knife to her and slashed her body open—just like Jack the Ripper—and then stuffed her into the wooden trunk, breaking her leg to fit her in. Why had he remained in Dundee, living alongside her body for around five days before finally giving himself up? Was it remorse that took him to the police station that night? Was it a sudden sense of responsibility? Given what was known of the character of William Bury, neither seemed likely.

The trial was set for 28 March, when the spring court circuit was due to sit in Dundee with Lord Young presiding. All the evidence was to be heard in a single day as the judge did not trust the Dundee public not to interfere with the jury members. The city was vehemently opposed to the death penalty.

The police must have been pretty convinced that this was an open-and-shut case, but they had not reckoned on the courtroom skills of the two other doctors brought in for the defence, Drs Lennox and Kinnear. They were required only to provide grounds for reasonable doubt, not to prove innocence. Bury was pleading not guilty, and it was the Crown's job to make its case.

The defence doctors did not dispute Bury's assertion that he and Ellen had been drinking, although they did note that there was no smell of alcohol from her stomach contents. For whatever reason—and they weren't obliged to provide one—they offered the hypothesis that Ellen could have self-strangulated using the cord found round her neck, with the assistance of partial suspension, perhaps from the door handle. Maybe Bury picked up the knife to cut her down and, distraught and not in his right mind, took the knife to her abdomen in the process. But they believed this took place postmortem as they found no lividity associated with the soft tissue around the wounds. Afterwards Bury panicked and bundled her into the trunk. He then spent five days agonizing over what he had done until, unable to live with it any longer, handed himself in to the police. Nobody was able to explain why he burned all Ellen's clothes, why he had her jewellery in his pockets when he went to the police station or why the floor of the flat had been cleaned while the rope and the bloody knife were left in clear view. So much of the case simply did not add up. (If you are interested in the full story, do read Euan Macpherson's book *The Trial of Jack the Ripper: The Case of William Bury*.) In his summation, the judge told the jury they had only two things to consider: was this murder or was it suicide? The fifteen men returned with a guilty verdict, but with a recommendation for mercy, reflecting the local distaste for the death penalty. When Lord Young asked them for their reasons, they said it was on the grounds of conflicting medical evidence.

Irritated, Young sent them back to the jury room until they were agreed on a clear verdict. It took them only five minutes to come back with a unanimous decision: guilty. Lord Young passed the mandatory death sentence for murder and, four weeks later, some time between 8 a.m. and 9 a.m. on 24 April 1889, William Bury was hanged by the neck until he was dead. It was at least a "good" hanging: he died almost instantaneously due to the fracturing of the second cervical vertebra. He was twenty-nine years old.

On 7 January 1889, before William and Ellen Bury set foot in Dundee, the first Cox chair of anatomy at Dundee university, Professor Andrew Melville Paterson, gave his inaugural lecture. Obtaining bodies for dissection remained difficult in areas of the country where hangings were infrequent.

In the wake of the nefarious activities earlier in the century of Burke and Hare—who resorted to murder to supply cadavers to anatomists in Edinburgh, one of Europe's leading centres of anatomical study—the Anatomy Act of 1832 had acknowledged the need for bodies for scientific purposes and granted licences to anatomists, giving them legal access to unclaimed corpses. For the first time it also allowed for a person's next of kin to donate their body to medical science. But the majority of cadavers were still sourced from those who had died in prison, hospitals, asylums or orphanages, or who had committed suicide.

Three months after Paterson took the chair, the body of William Bury would have been legally available to Dundee's licensed anatomists and Paterson would surely not have missed the opportunity to acquire it. Unfortunately, although the university has records of corpses acquired both before and after Bury's death, there is nothing to confirm whether his body ever entered the anatomy department. However, we do know for certain that Paterson had some connection to Bury's remains, because seven cervical vertebrae that used to sit on my desk in Dundee, including the C2 vertebra that showed the classic hangman's fracture, were catalogued in the museum collection as having belonged to William Bury. Perhaps Paterson removed the neck vertebrae himself; perhaps either Templeman or Stalker did it for him. We just do not know.

To mark the 130th anniversary of the Cox chair of anatomy at Dundee university, we decided to reconsider the evidence presented at the trial of William Bury and act it out in court—indeed, the very same Sheriff Court in which he had stood trial just a couple of months after the inauguration of the academic chair. The medical reports and the notes taken by Lord Young are stored in the Scottish records office in Edinburgh and we had full access to them. We decided to present only the medical evidence, as this had been the basis of the original instruction from Lord Young, but to do so in the light of contemporary forensic knowledge. It would be fascinating to test whether modern-day science would uphold the views of the jury or overturn their verdict.

A serving supreme court judge, Lord Hugh Matthews, sat on the bench—not in his robes, as that would have been improper: this was not a retrial, but public engagement with science. The legal team for the prosecution were the Mooting Society from the University of Dundee, and these incredibly lucky law students were coached by

Scotland's principal Crown counsel, Alex Prentice QC. They would present testimony from one witness only: a highly regarded forensic pathologist, Dr John Clark, who would confine himself to the evidence recorded in the reports of Drs Templeman and Stalker, the original Crown pathologists.

The legal team for the defence were from the Mooting Society of the University of Aberdeen, coached by Dorothy Bain QC, a leading Scottish lawyer who had been a Crown office advocate depute for many years. They, too, would call only one witness: Dr Richard Shepherd, an equally renowned forensic pathologist, who would work with the evidence provided by Drs Kinnear and Lennox, retained by the defence. The fifteen jurors (fifteen is typical in Scotland) were selected at random from members of the Dundee public following a campaign run by the local newspaper, the only difference being that, to reflect societal as well as scientific advances, we did not restrict ourselves to outmoded nineteenth-century financial or gender eligibility criteria, and the jury was therefore made up of both men and women. The event was filmed by Dan Snow's company and continues to be used today in the instruction of law students at both Dundee and Aberdeen universities.

The Sheriff Court was packed to capacity as the young man appointed to stand in for William Bury took his place in the dock, flanked by a police officer in period costume. It was Dr Clark who provided the defence team with the snippet they had been looking for. He offered the view that self-strangulation could not be ruled out, that it could have been achieved from a low-level object and that he could not discount the possibility that this could have been the door handle. This gave the defence all the ammunition they needed to champion the cause of reasonable doubt. But would the jury buy it?

After all the evidence-in-chief had been heard and all the cross-examinations and re-examinations completed, the case was summed up by the judge. His most memorable direction to the jury was: "Normally I would say there is no pressure of time, but in this case, I give you fifteen minutes.' It was strange to hear laughter in the courtroom, especially in one considering something as serious as murder and dismemberment, and after everyone had taken it all very seriously and played their parts perfectly.

The decision the jury had to make was whether they agreed with the prosecution or whether the defence had introduced reasonable doubt: as this was not a retrial, it would have been inappropriate for them to pronounce on guilt or innocence. When they returned, they were split 13–2, on this occasion, in favour of the defence. They did not think that there was sufficient evidence to convict for murder, again due to the conflicting medical evidence. The twenty-first-century jury was largely upholding the first instincts of their predecessors 130 years before.

However, William Bury was not to be let off. The judge, who could not alter the verdict, informed him: "Mr Bury, please stand. There is good news and there is bad. This jury has found you not guilty, but I think you did it, so you are still going to hang. Take him away."

So Bury was once again led from the courtroom and down the steps to the cells, only this time, happily, he was not really going to be hanged by the neck until he was dead and nobody was going to dissect his neck to remove his bones.

There was a little twist to the tale that few were aware of. In the dock that day alongside the modern-day William Bury were the mortal remains of the real defendant. I had brought his cervical vertebrae back into the same courtroom where he had stood trial on 28 March 1889. And so, on 3 February 2018, part of his body was present as the jury returned a very different verdict.

What do I think would be the outcome if Bury was put on trial today? I reckon he would have got "not proven," known colloquially in Scotland as the "bastard verdict," and would therefore have been acquitted. This verdict basically indicates that the jury believes the defendant may be guilty but that there is insufficient evidence to prove it. While history records what happened to Bury, and is supported by the evidence of his vertebrae, we do not know whether Ellen was strangled or strangled herself. Even if we could examine her remains today, we would not be able to tell the difference.

I was pleased with my quirky choice of keepsake for everyone who took part in the event. They each received a 3D-printed replica of William Bury's C2 vertebra, complete with its classic hangman's fracture, in a presentation box. I only recently found out that Dan Snow gave his replica vertebra, in its beautiful box, to his wife as a Valentine's gift. And they say romance is dead.

The vertebrae, then, tend not to bear any trace of a soft-tissue injury such as strangulation, whereas the upper cervical vertebrae may speak to death by hanging. What about cases of decapitation? In this situation, the vertebrae will probably carry evidence of the event if the decapitation, or attempted decapitation, has been intentional. This usually results in the first three cervical vertebrae staying with the head and the last two with the chest, so the implements used by an assailant are most likely to leave their mark between C3 and C6, that is, on C4 or C5.

Deliberate separation of the head from the body tends to take place in the same area of the neck whether the head is being removed from the front or the back. From the front, it is below the level of the mandible, which can get in the way of chopping or sawing. From the back, it will generally be about halfway down the long stretch of the neck. Any marks seen higher or lower than this would be viewed as out of the ordinary (not, of course, that any form of decapitation is ever ordinary). For example, in the infamous "jigsaw murder" detailed in *All That Remains*, in which the victim's body was dismembered and scattered across two counties, the cut made to sever the head was clean and very low down on the neck. But this turned out to have been a professional job: it emerged during the trial of the perpetrator that he was a skilled underworld "cutter" who specialized in dismembering bodies.

In one case of alleged murder and dismemberment, my team was asked by the defence to consider the forensic reports offered by the Crown scientists. Much of the time, forensic anthropologists are engaged by the Crown, but of course, we do also appear for the defence. Whichever side has retained your services, it is important that the evidence you give is the same, because ultimately, you are a witness for the court and not for one side or the other.

The crime came to light after a family walking through woodland with their dog came across an apparently abandoned trainer with a sock inside it. On taking a closer look, they saw, to their consternation, that the sock contained foot bones. The police were called, it was established that the bones were human, and further remains were finally located hidden among the roots of a fallen tree. It was clear that over some time the body had been predated by foxes and their cubs and the bones had been scattered around the area. Hands and feet tend to be taken first by animals as they are both accessible and portable. Predators will then chew on bigger bones in situ. It is quite common for the skull, being such a heavy and unwieldy object, to be left where it is, as it was with these remains.

What other body parts could be found were brought together, and DNA confirmed that they were those of Jamal, a middle-aged man who had been reported missing some three years earlier. There was no longer enough evidence to provide an obvious cause of death, but a forensic anthropologist was asked to examine the remains for the Crown and concluded, from marks found on the vertebrae, that the victim had been decapitated.

Jamal had inherited some money from his late mother's estate and because he had learning difficulties, he trusted his daughter's partner to look after his bank account. Unbeknown to him, while he was constantly kept short of cash, his money was being splashed out at antiques fairs and on expensive cruises. Within two months there was only 78p left. It seems likely that he eventually realized what was going on and confronted his daughter's partner. The police believed that this was probably what escalated an altercation into Jamal's death.

His daughter's partner was charged with aggravated murder. The "aggravated" element related to the concealment of, and disruption to, the body, an offence viewed by the judicial system as an additional degradation. The question of whether the body had been dismembered by the murderer was important because, if found guilty of the aggravated charge, the accused could find himself sentenced to a whole life tariff without possibility of parole.

Professor Lucina Hackman and I examined photographs from the scene and the postmortem, and those taken by our Crown colleagues during the course of their inquiry. It is a requirement of our profession that we also keep detailed notes that will permit any other scientist to replicate our investigation and come to their own opinion.

Lucina and I were called to appear at the trial as expert witnesses for the defence. It all started well. However, once the Crown expert began to give evidence, it became clear that there was going to be some dispute with regards to the allegation of decapitation. We already had serious reservations about opinions our counterparts had reached on this matter.

Sometimes we are in total agreement with the experts called for the other side. On other occasions, we may believe a scientist has strayed beyond their area of expertise or has given undue weight to evidence that could be interpreted very differently. When appearing for the defence, you are asking yourself every step of the way whether there could be another sound explanation for the conclusion arrived at by the Crown. It is about reasonable doubt, and that is relevant for both the Crown and the defence.

One of the interesting things you can do in an English court, which you cannot do in a Scottish court, is sit in the courtroom, listen to the live testimony of the experts for the opposing counsel and transfer information to your legal team as evidence is given and questions arise. This enables you to indicate where you are in agreement and which aspects you believe need to be challenged. The detail might not be in the written report but may emerge when the reasoning becomes clearer under legal interrogation.

For us, in this case the evidence did not stack up. First of all, the head was found with the body, which was odd. Why would someone go to the trouble of cutting off the head if they were just going to hide it with the rest of the body? There was no evidence of deliberate removal of any other body parts, and in a criminal dismemberment, for which by far the most common motive is easier disposal and concealment of the corpse, or to prevent identification, it is upper and lower limbs that tend to be removed in preference to a head.

Secondly, the "cut" marks referred to by the Crown's expert were on the second cervical vertebra, which would be unusually high for a successful decapitation, or even an attempt at it. If the cut were made from the front of the neck it would have been extremely difficult to accomplish. The mandible would be in the way of whatever tool was being used and it would necessitate deep dissection down through the soft tissues, which would be very messy. There was no evidence of this. Thirdly, the cut marks themselves didn't look like those that would be made by any saw, knife or cleaver that we had ever seen, and no murder weapon or dismemberment implement had been found.

Straight away the Crown witness introduced an opinion that had not been recorded in her report. This is not permissible and so the defence instantly objected. The judge decided to give her some leeway and asked us if we would be prepared to meet her outside the courtroom to establish whether we agreed with this new evidence.

This we did, and we did not agree. In court the following morning, the Crown expert changed her mind again and the judge lost his patience. Lucina and I sat there not knowing where to look as he gave her a dressing-down and then adjourned the court. We tried to put on our best poker faces but it was a challenge. The defence barrister then told us that our evidence was no longer required and we could go. It seemed that the part of the charge relating to decapitation was going to be dropped as a result of the problems with the Crown's expert and her testimony, rather than anything to do with the evidence per se.

We had spent two days in court and had not offered a word from the witness box. The Crown's case for dismemberment had collapsed purely through the inexperience of their scientific witness. This is a story we have since used to hammer home to our students and trainees the importance of interpretation of evidence and understanding the processes of legal trial. If you fall foul of these, you may never get as far as giving your expert opinion.

In the end, we were to learn from press reports of the trial that the defendant had been found guilty of murder and sentenced to at least nineteen years in prison before being considered for parole. He was not, as we had been informed, charged with aggravated murder.

But what about the "cut" marks on the cervical vertebra? If they were not made by a tool, what could have caused them? To address questions like these, and in the analysis of human remains in general, a forensic anthropologist must not only think anatomically before they think forensically, they need expertise and experience beyond bone anatomy.

When a body decomposes, the first cervical vertebra normally remains associated with the skull because of the strong ligaments that bind them together. If the skull eventually rolls away or is removed by animals, this often leaves the C2 vertebra as the most exposed part of the column. We believed that this is what could have happened in this case, and that the marks were in fact the drag marks of canine teeth across the surface of the bone.

A knife makes a clean cut, with straight sides that mirror the width of the blade and a floor that may take on the shape of the knife or saw that caused it. It was clear from the photographs that these marks were more scores than cuts, with no depth to them. This did not rule out the use of a tool—they were consistent with tentative cuts, perhaps made by someone uncertain about how to remove a head, and with the "chatter" marks we see when a blade has skittered across the surface of a bone without "biting." However, we could also see paired triangular punctures into the bone that were not consistent with a blade of any kind. What they did match with perfectly, as did the line of "chatter marks," was the average distance between the upper canines of an adult fox.

So there was no decapitation or dismemberment, just an over-enthusiastic Crown anthropologist who had unwittingly led the investigation down a blind alley. It is of course essential that justice is done—a man was cruelly murdered, and his murderer is being punished for that—but the defendant is entitled to be judged on the evidence, and to a penalty that is fair and appropriate to the crime. They should not be found guilty of something they did not do.

The victim in that case was at least eventually found so that his

remains could be named and laid to rest. What leaves forensic anthropologists with an enduring uneasiness are those disappearances and
deaths that are not solved, either because all efforts to find a body
prove fruitless, or because we have a body we cannot name, which in
some ways feels even worse. The knowledge that we have done everything we possibly can does not alleviate the sense of a job unfinished
and justice unfulfilled.

In such cases there is usually little doubt that some kind of crime
has occurred, but the answers to precisely what, who is responsible,
and sometimes who the victim might be, prove elusive. It could be
argued that these should be classified as perfect crimes, since both
perpetrator and victim remain unknown. Or almost perfect crimes: I
suppose the truly perfect crime is one that nobody realizes has ever
happened in the first place.

There is some irony in the fact that, once a name has been assigned
to a body, our natural inclination is to anonymize the victim out of
respect for them and their family. But while a body remains unidentified we do the opposite: we publicly spread all the information we
have in the hope that a name might one day be forthcoming.

The challenge is often tougher when a victim either originates
from, or has chosen to live within, a transient or chaotic community.
The "Angel of the Meadow" was one such victim. Here, the vertebrae
were able to tell us something about her, and about what had happened to her, but not to lead us either to her identity or that of her
killer.

The murder came to light when skeletal remains were found
during redevelopment of a building while a mechanical digger was
shifting soil. The skull was spotted first and then the rest of the bones
were discovered under some sections of carpet. It was believed, from
the clothing, that the body had probably been there for thirty or forty
years, since the 1970s or 1980s. The anthropologically trained SOCO
had identified that this was a female, between eighteen and thirty years
of age when she died, of average height and most probably Caucasian
(which would not rule out someone from the Indian subcontinent, the
Middle East or north of Saharan Africa). It is likely that she was naked

from the waist down when her body was concealed here, as a pair of tights, an empty handbag and a single shoe were found nearby.

At Dundee we were asked to examine the remains for clues that might help to identify her, and also to try to make some sense of how the fracturing of her vertebral column might have occurred. In a violent death, we are used to seeing fractures of the face, mostly the nose, cheek, jaw or teeth, or in the neurocranium, caused by blunt-force trauma, but the level of specific fracturing found in this woman's neck was unusual.

The lower part of the first cervical vertebra showed a self-explanatory "starburst" fracture of the joint surface, between C1 and C2 on the right-hand side. This was a localized crushing injury which had not substantially affected the second cervical vertebra. However, further significant crushing injuries were also detected on the left side of C3. In summary, we had a crush or compression fracture on one side of C1, no evidence of injury on C2, but more crush fracturing on the opposite side of C3. The question was, how could these odd injuries be translated into a plausible cause and effect?

It was possible that the cause may have been one of those hideous methods of "dispatch" seen in violent spy movies, where the head and jaw are grasped between two hands and rotated violently. In this instance, the head could have been turned hard to the right, with hyperflexion of the neck (the neck being bent a long way forwards) resulting in dislocation between C1 and C2, perhaps severing or crushing the spinal cord. The damage to C3 might have been inflicted by the force of the severe rotation. What was certain was that her death was brutal, if perhaps mercifully swift.

But who was she? She'd had several fillings and other dental work, so records must have existed for her at a dental surgery somewhere. Three women had been reported missing in the area around that time but their dental records did not match the teeth or dental work of our unidentified skeleton. DNA analysis did not help, either. We were able to produce a reconstruction of her face, which generated several leads that resulted in investigations in a number of places around the world, from Tanzania and the USA to Ireland and the Netherlands. Yet to this day the Angel of the Meadow remains unidentified and is buried in an unmarked grave.

Is there anyone still around who thinks about her and wonders what happened to her and where she is now? Is the person who did this to her still alive? Do they live with daily guilt and a burdened conscience? We can only hope so.

Such an injury is not something that frequently comes the way of a forensic anthropologist, but when all that is left is bones, sometimes an anthropologist is the best place to go for answers.

When the upper part of a human torso was washed up on a beach at Southsea in Hampshire, the local police force called me for help with trying to establish what had happened to the victim. The body had not been in the water for very long and was still relatively fresh, which must have been a shock for the students who found it. Then the pelvis was discovered, and shortly afterwards the legs washed up on a different part of the beach. Two days later, a man called the police in a neighbouring county, concerned that he felt he had done something wrong, but said he couldn't remember what it was. He was met at the station by a police constable who noted that he was dirty and dishevelled and seemed confused. He was known locally to be a drinker and perhaps abused other substances too.

The police went with him to his flat and found nothing immediately suspicious. But once the remains were identified as those of a friend of his, he was arrested for murder and dismemberment, which he denied. The deceased was a man of low IQ who lived in a camper van and the accused allowed him to wash at his flat in return for feeding his cat, Tinker, from time to time. Nobody knows what happened, whether there was an argument, or perhaps a scuffle broke out after too much alcohol was consumed. The accused was reported to have been abusive towards his friend on occasion. Perhaps there was a knife involved. Whatever the case, the alleged killer was left with a body and the problem of disposing of it.

Most dismemberments result in a body being divided into five or six body parts with the torso generally remaining intact. Attempting to cut up a torso makes a terrible mess unless you remove all the internal organs first. In this case, the viscera had been removed, the torso split across the lumbar vertebrae and the pieces wrapped in bin liners and a pink shower curtain. The external genitalia had been cut off and the head, arms and viscera have never been found. The accused owned a bicycle with a butcher's box on the front and the police believed that this was what he used to transport the body parts to the seashore, where he threw them into the water.

My team was asked to examine the sites of dismemberment and to look for additional cut marks on bones that might explain how the body had been taken apart. The police brought some of the remains up to us in Dundee as we had the facilities to process them in a way that would preserve the cut marks better than the usual method, which involves simmering in warm water with a biological detergent. While the body had been relatively fresh at its first postmortem, by the time these pieces reached us they were starting to become a little more antisocial.

At Dundee we had a colony of dermestid (flesh-eating) beetles. These occur naturally in the soil and are one of the first waves of insects whose activities assist decomposition, slowly rendering a body down to a skeleton. Most of the time we fed them on their favourite delicacies, mice and rabbits (they did not like marine animals, so no fish or seal for these little creatures). We used to get a few complaints from our university colleagues about their somewhat distinctive aroma but they were an incredibly useful resource for safe and gentle defleshing. We placed the material we had in with the beetles and returned frequently over the next few days to remove each section when it had been completely cleaned.

Once we were able to see the bones clearly, it was evident that the shoulders had been disjointed with a sharp blade and that a similar blade had been used to remove the head and to separate the lower limbs from the pelvis. However, the sternum and the lumbar vertebrae had been dealt with very differently. They showed the regular striations that are left behind by a power saw. This saw had been used to unzip the sternum and gain access to the chest to remove the internal organs, including the heart and lungs. The upper torso had been separated from the pelvis by the same power tool, which had cut across the fourth lumbar vertebra, presumably after the evisceration had been performed. We were also able to find small cut marks on the sides of the lumbar vertebrae that could have been evidence of the removal of the internal organs with a knife.

For some reason the pathologist had told the police that the cut across the lumbar vertebrae had been effected by a Japanese wood saw, and that set them off on a wild-goose chase, first of all to establish what one of these actually was, and then to figure out why the accused would have such a specialized tool and where it might be found. As he was a scrap-metal dealer, it was more likely that he would have access to a power saw than to a Japanese wood saw. Lucina did share this thought with the police, but if the pathologist says it is so, then it must be so . . . And yet they never found a Japanese wood saw.

How, then, was the accused linked to the death of his friend? Well, to start with, some blood was found in his flat, but there was another fascinating strand of evidence that broke new ground for forensic science. Cat hairs recovered from the shower curtain in which the torso was wrapped were sent to the US for analysis of the feline mitochondrial DNA, which is passed down through the maternal line. The results prompted further testing in the UK at Leicester University's department of genetics, which established that there was only a 1 in 100 chance that the hairs did not belong to Tinker, the defendant's cat.

This was the first time cat DNA had been used in a criminal trial in the UK. Due to domestication, cats have fewer genetic variants than humans, so it was fortunate that Tinker's genetic make-up was relatively uncommon, but as work continues to make the tests more specific, analysis of animal hair could become a very useful source of evidence in the future.

The jury acquitted the defendant of murder, but the cat hair, supported by further fibre analysis of the shower curtain, which was from his home, helped to convict him of manslaughter. He was given a life sentence and ordered to serve a minimum of twelve years in prison.

As for the part we played, our evidence on the manner of dismemberment was accepted by the court, and we were therefore not required to attend to testify, which is always a blessing. But Lucina and I have lost count of the number of police officers who have said to us, "Oh, was that the Japanese wood saw case?" Every time that blooming Japanese wood saw is mentioned we roll our eyes.

Each of the vertebrae can tell us something of the age, sex or height of an individual and shine a light on pathology, disease and injury. But perhaps their greatest value to forensic anthropology lies in the information they convey about the trauma and damage inflicted on those who become the victims of violence, before, during or after their death.

4

# The Chest
## *Thorax*

*"Cut open my sternum and pull my little ribs around you"*
Purity Ring
Pop Band

The bony walls of the thorax have several functions but their primary purpose is to protect the delicate lungs and heart and to provide a structure for muscle attachment, particularly to help with breathing but also with movement of the upper limbs. The job requires thirty-nine separate bones: twelve pairs of ribs, anchored at the front to the sternum (which consists of three bony parts) and at the back to the twelve thoracic vertebrae.

Given that the organs they protect are essential to life, it is perhaps to be expected that, along with the skull, the thorax is the area of the skeleton that is the most common focus for violent assaults. If you want to kill someone swiftly, aiming for the head enables you to attack the brain, but its surface area is relatively small and as the skull is thick in parts, it can resist damage more readily than the comparatively fragile bones of the chest. With its much larger surface area, the chest offers a bigger target and access to the heart, as well as to some very large and unforgiving blood vessels. So the thorax tends to be the most frequent choice for inflicting injury using a variety of weapons and methods: sharp-force trauma (for example, by stabbing), blunt-force trauma (such as kicking), and ballistic trauma (shooting).

As well as being easier to fracture, the bones of the thoracic region provide convenient gaps through which sharp implements can easily be inserted. This was apparently the fate of Richard Huckle, the UK's most prolific paedophile.

A vile predator who posed as a devout Christian, twenty-eight-year-old Huckle was alleged to have abused over twenty-three children, in an age range of six months to twelve years, between 2006 and his arrest in 2014. Most of his victims were Malaysian children who lived in the capital, Kuala Lumpur, but it is possible that he may have committed further crimes in the UK and elsewhere.

The depth of his depravity knew no limits. He had been compiling what was essentially a "how to" manual for paedophiles, which he was about to publish on the dark web, the internet underworld beyond the reach of normal search engines. Entitled "Paedophiles and poverty: child lover guide," his treatise spelled out in detail how to groom and gain the trust of young children from impoverished backgrounds. These children frequently had nobody to care for them and their loyalty and dependency could be bought by kindness, small amounts of money and cheap gifts. In such environments, it doesn't take much to persuade a child to acquiesce to the wishes of a deviant. UK police, alerted by a specialist Australian child abuse unit that Huckle was due to return to England to spend Christmas with his family, arrested him as he stepped off the plane at Heathrow airport.

This abhorrent case landed on my desk in 2015, after Huckle was charged with ninety-one counts of indecent acts against children. We were given the task of looking at nineteen still images and nearly eight minutes' worth of video to determine whether the same offender was shown in them all and whether Huckle could be excluded as that offender.

Still images are relatively straightforward to scrutinize as they encapsulate a static moment. Videos are more difficult because you are exposed to the changing gestures, movements and facial expressions of both victims and perpetrator. Eight minutes of video might not sound like very much, but for the purposes of examination it has to be broken down into single frames, and as there may be multiple frames to every second of film, you can very quickly find yourself dealing with over 50,000 separate images. And when these show abuse of a child, eight minutes is interminable.

We were able to confirm that it was highly likely to be the same man in all of the images and that, on the basis of various anatomical features visible in his hands, genitals and lower limbs, the man was most likely Richard Huckle. This was supported by the superficial vein patterns on the backs of his hands and on his penis, the areas of punctate pigmentation (moles) on his hands, forearms, thighs and knees and the pattern of skin creases on the knuckles of his thumbs, fingers and palms. It was clear that Huckle had a condition called phimosis, which occurs in about 1 per cent of non-circumcised males, where the foreskin is too tight and cannot be retracted around the tip of the penis. Most adults with this condition opt for surgical intervention to release the sphincter. Huckle had not, which further reduced the likelihood that the perpetrator might be someone else.

The police advised him that our evidence was a strong reason to change his plea. He eventually agreed and pleaded guilty to seventy-one of the charges. He was given twenty-two life sentences and a minimum prison term of twenty-five years before he would be considered for parole. Huckle was into his third year at HMP Full Sutton when it was reported that a fellow inmate strangled him with a bandage-like ligature and then stabbed him to death with what was described in the media as "a makeshift weapon, most likely fashioned from a sharpened toothbrush."

In the hands of those who know where to place it, almost any innocent household object can be turned into a lethal weapon. Something as simple as a filed-down toothbrush, forcibly rammed into the space between the fifth and sixth ribs on the left-hand side, just below the nipple, will enter the heart, which lies directly behind the sternum and the front ends of those ribs. The punctured heart will pump its blood into the body cavity and life is extinguished. So just one stab can kill and, when the implement is made of plastic, it will probably leave no visible evidence of the event on the bones.

My reaction to Huckle's murder was complicated. I had been

reassured that his change of plea was perhaps a sign that he had accepted responsibility for his crimes (although he may simply have realized that he was cornered and had nowhere else to go) and I felt the length of his sentence was appropriate. It took him off the streets and he would have been fifty-three years old before he was even considered for parole, which would have been plenty of time for efforts to be made to rehabilitate him.

I am an optimist. I want to regret that a young man of thirty-three met a violent end, but I find it hard to summon the necessary compassion for a person who did so much damage to so many vulnerable children. While I am disappointed in myself for my inability to be more forgiving, I suspect I would not even be countenancing the option of forgiveness if his victims had been my children or grandchildren. Do I believe in the death penalty? No, not really, but the case of someone like Huckle probably comes as close as anything could to persuading me to change my mind.

Huckle's killer knew where to strike. It is much more difficult to stab into the chest, even with a more effective weapon, if you aim for the vertical strip in the middle. The sternum, the hard, bony breastplate at the front of the chest, is made up of three parts. The old anatomists decided rather fancifully that it looked a bit like a sword, with a broad handle, or pommel, at the top, a long, thin blade in the middle and a sharp pointy bit at the bottom. So the top section of the sternum, the "handle," is known as the manubrium (from the Latin *manus*, meaning hand). The middle area, the body, or mesosternum, is sometimes referred to as the gladiolus, like the flower, which takes its name from the Latin for sword, as does the word "gladiator." The terminal point is the xiphoid process (in this instance from the Greek, meaning sword-like).

If you picture a chicken, a creature whose anatomy is familiar to most of us, the equivalent of the sternum in its skeleton is the keel we see in the midline between the two breasts. Our sternum lies directly under the skin of the chest and has no covering of fat or muscle.

No matter how obese we may become, it remains palpable, which means that being struck in the sternum is really painful. Fractures are common. After the use of seatbelts in cars was made compulsory in 1989, the number of sternal fractures caused by drivers bouncing off the steering wheel reduced markedly, but such fractures are still seen in sports-related injuries.

Its closeness to the surface of our bodies makes the sternum a useful landmark for first-aiders or point of access for doctors. It is a handy target for cardiopulmonary resuscitation (CPR) as it offers a solid base on which to pump when trying to kick-start a dicky ticker. Pressure on the xiphoid process, however, should be avoided as it can fracture, and when this happens it has been known to perforate the liver and cause a fatal haemorrhage.

The sternum is also a convenient site for biopsy when doctors wish to aspirate bone marrow. Evidence of surgical excision through the bone to gain access to the chest, for example, during open-heart surgery, can be a rather obvious clue to the medical history of an individual. Anatomists frequently see the handiwork of the cardiothoracic surgeon in the dissecting room in the remains of elderly donors who have bequeathed their bodies to further our knowledge of the human form and to train our students. Many chests bear witness to the cutting and mending that takes place in emergency surgery when there is no time for careful planning or non-invasive approaches.

Neither the xiphoid nor the mesosternum tend to survive long after burial, as the layer of covering compact bone is thin. But the manubrium usually endures well, particularly the upper part, where it is reinforced to take the strain of the joints with the clavicles, or collar bones, on either side.

The manubrium can be very useful in age determination in young people, thanks to the small, thin flakes of bone that fuse to the joint surfaces in the early teenage years to complete the growth of the joints. It is an area that is overlooked by almost every other profession concerned with forensic science, but one that an experienced anthropologist will always check.

There are quite a few developmental anomalies that manifest themselves in the sternum which can prove helpful for identification purposes. During its formation, the sternum may remain perforated in the midline, leaving a hole reminiscent of a bullet wound, which can be misleading for the rookie. It causes no clinical symptoms—it is simply a result of defective fusion when the bone was growing. Anatomists love to use such specimens in student exams. They bring forth some extravagantly imaginative descriptions of violent homicide and ballistic trauma but in fact, if less thrillingly, they are just a normal variant.

Sometimes the xiphoid can become quite long, and even bifurcated, so that strange lumps and bumps become noticeable on the upper midline abdomen with age, which can cause unnecessary alarm to its owner, who may fear they have a tumour. It can be quite tricky to identify a bony xiphoid process in isolation and sometimes this is only achieved by accounting for everything else. If all the other bones are present and we are left with an odd, pointy strip of bone, we can be pretty sure it is the xiphoid of a middle-aged or elderly individual, usually male.

Pectus carinatum, or pigeon chest, occurs when the cartilages associated with the ribs overgrow, producing a "keel-like' protrusion of the chest wall. This can be caused by a number of conditions, including rickets (a result of vitamin D deficiency). Pectus excavatum, or sunken chest (known by the ever-jocular medics as "pirate's treasure"), is its anatomical opposite. This can impact on the normal functioning of the heart and the lungs. Its cause is uncertain. It may be simply a congenital defect in the formation of the sternum. Sometimes when the sternum does not develop normally, the heart can grow outside the fetal chest. This requires some tricky intrauterine surgery to open the sternum of the fetus and pop the heart back where it belongs so that it can continue to grow as it should.

All in all, the tri-osseous strip of the sternum is very useful to the medical profession, but it does not tend to excite too many forensic experts—except for anthropologists, who are always on the alert for unusual-looking chest bones as these may indicate a particular developmental condition that might provide an important clue to the identity of an individual.

It was the sternum that enabled us to significantly narrow down the age of Jin Hyo Jung, the South Korean woman found in a suitcase in the investigation recounted in Chapter 2. The key was the very specific age changes that occur in this bone. In a child, the sternum is often in six separate pieces which start to fuse together in the midline as the child grows until they ultimately form the typical adult three-piece structure by the late teenage years. Further changes continue throughout puberty and into the early twenties at the sides of the bone, at the sites where the cartilages of the ribs articulate with the sternum: here, delicate slivers of bone appear in the region of the cup-shaped joints, fusing first in the upper and finally in the lower borders of the sternum.

We could see on X-rays that our as yet unidentified young Asian woman had these shards of bone nestled in the little cups of the joints, from which we calculated that she had to be under twenty-five but older than twenty when she died. She was in fact in her twenty-second year.

The sternum is a pretty good indicator of sex, too. It tends to be longer in males than in females and is generally larger and more robust. If you have better-developed muscles (particularly the pectoralis major, or pecs) on the front of the chest, then it follows that they are going to need to be attached to bigger, stronger bones. Of course, having a bigger, stronger sternum does not always mean you are male—think female weightlifter, shot-putter or javelin-thrower.

The bars of cartilage that join the bony ribs to the sternum are called costal cartilages ("costa" as in rib, and nothing to do with coffee). The cartilages are the remnants of the precursors of the ribs which haven't yet got round to turning into bone but retain the ability to do so with advancing age—a process known as ossification. The first signs of this can be visible as early as the late teens or early twenties, and it becomes progressively more developed as we get older, to the point where almost the whole strip of cartilage may eventually be replaced by bony bits.

Sometimes the edges of the sternum send tongues of bone out into the upper and lower borders of the costal cartilages, resulting in a structure that looks like a spider, with the sternum at the centre and the bony cartilages extending out on both sides like legs. This sternum–cartilage–rib combo is for this reason sometimes referred to by older anatomists as an arachnid. Its official designation is a plastron, a word for breastplate which has various other definitions, ranging from a fencer's chest-protector and a nineteenth-century ornamental woman's bodice to the name for the underside of a turtle.

We always suggest an X-ray of the plastron if at all possible, because you just never know what it might tell you. And one thing is certain: if you don't look, you won't find anything.

The costal cartilages were strikingly informative in a case where skeletonized and partially scattered human remains were discovered in woodland on the outskirts of a small Scottish city. The deceased was wearing one high-heeled shoe, size 8, and had possibly been naked from the waist down as there was no evidence of any remnants of clothing that may have been on the lower half of the body, although there was a bra and a blouse associated with the upper part. Other items of female paraphernalia were retrieved during a search of the area, including a plastic handbag containing make-up and a handkerchief, but no money or credit cards.

There is a tendency to be swayed initially by circumstantial evidence when a body is found—particularly when it is skeletonized. The recovery of female clothing and a handbag quite understandably suggested to investigators that this was most likely to be the body of a woman. However, such linear thinking can take an inquiry in completely the wrong direction if we are not careful. Assumption is the mother of all mistakes.

As I started to examine the skeleton with this bias in mind, I quickly began to get very confused. The skull appeared to be more masculine than feminine, and so did the pelvis.

When investigating skeletal remains, anthropologists usually start with sex determination, as that tends to be the easiest aspect of identity to establish with reasonable accuracy and it automatically rules out missing persons of the opposite sex. It is not uncommon for bones to display traits that are neither strongly masculine nor strongly feminine, but when the conflicting information is coming from the skull and the pelvis it is troubling, as these bones normally give us our best opportunity to get it right. I set sex determination to one side for the moment and turned to age estimation. I was much happier with the accuracy of my assessment here. The woman was somewhere between thirty-five and forty-five years old, most likely at the lower end of that range.

It was when I asked for a routine X-ray of the chest plate that the mists began to part. The way in which the costal cartilages start to ossify is dictated by the prevalence of either the male hormone testosterone or the female hormone oestrogen circulating in the vascular system. In ageing males, the bone in the cartilages is laid down along the upper and lower borders of the cartilage bar and can eventually fuse on to the front end of the rib, resulting in rib ends that look a bit like crab claws. On an X-ray, this new bone mimics the structure of the rib, with a thicker outer shell and a honeycomb appearance on the inside. This is the effect of testosterone on ossification of hyaline cartilage, which forms the costal cartilages.

If the dominant hormone is oestrogen, bone is laid down very differently in the costal cartilages. We will see dense, sclerotic nodules, mainly along the central core of the cartilage. So, from the cartilages alone, we might be able to proffer an opinion on sex with some confidence: do we have crab claws or a string of bony pearls? As ossification becomes more progressive with age, we are also able to provide a very broad age range (young, middle-aged or elderly) just from looking at an X-ray of the plastron.

So far, so good, but of course hormone levels can be changed artificially by medication or by disease. It is therefore logical that if you are biologically male but take regular doses of oestrogen, or biologically female taking regular doses of testosterone, the cartilages are likely to show both types of bone formation—crab claws and pearls. But before we get too excited about this, it is worth remembering that males naturally produce oestrogen and females testosterone. That means there is often a mix of both types of bone formation in both sexes. It is the proportion of one to the other that is important. In the costal cartilages of our woodland body, I could see quite extensive crab-claw ossification overlaid by dense, sclerotic nodules in the centre of the cartilages. Both were very pronounced. This required the kind of discussion that anthropologists need to have with each other first of all, when nobody else is listening. It is amazing how useful a toilet break can be when you want to talk to a colleague, out of earshot, to rehearse how to carefully phrase what you think you might want to say. What is said is not always what is heard, and what is offered as a theory has a habit of suddenly becoming gospel. Remember that Japanese wood saw.

After a sotto voce conversation with Lucina, I summoned the courage of our joint conviction and suggested to the police that the victim might be transgender. Given that they were wearing female clothes and that the skull and pelvis were so masculine, I believed it was possible we had someone transitioning from male to female.

While this probably wouldn't raise any eyebrows nowadays, twenty years ago it was quite a radical theory, and I suspect the police thought I was off my rocker. I was proposing that this person may have been born a biological male, had taken oestrogen supplements and had latterly been living as a female. The pathologist shrugged his shoulders and said that it was possible, but none of them seemed too convinced. However, once the DNA results came back, the presence of Y chromosome genes confirmed my early suspicions and the anthropologist's status was suddenly elevated from lunatic to miracle-worker. The victim was traced to a community that kept its distance from the police and consequently nobody had alerted the authorities that she was missing, if indeed anyone had noticed or cared. DNA from a relative confirmed her identity. Yvonne, whose birth name had been Martin, was a prostitute who worked the red-light district picking up gay men—a specialist known by the clients in those days, rather crudely and offensively, as a "chick with a dick." She had apparently been a heavy heroin-user and substance abuse had indeed been suggested by the evidence visible on her rib ends.

One of the most common clinical complications for heroin addicts is infection, which can manifest itself in the chest wall, where the junction between the ribs and their cartilages becomes inflamed. The most common culprit is *Pseudomonas aeruginosa*. We found evidence of previous infection at the front end of Yvonne's ribs but there was no clear indication of what had caused her death. Drug-related paraphernalia had been found all over the area where her body had been discovered, and it was known to be a place where addicts congregated to share needles. Had she taken an overdose, or shot up heroin from a bad batch? Perhaps that was what had happened, and her body had just been thrown into the undergrowth and forgotten. The bones in her chest told us some of her story, guided the police down the right track to find out who she was and allowed her to be buried with both her names, old and new.

Like the sternum, the ribs are susceptible to fracture because the layer of covering bone is relatively thin. Being curved, and connected at both front and back, they tend to snap either just in front of where they form a joint with the vertebra at the back (posterior angle), or towards the front (anterior angle), just behind where cartilage and rib meet.

At birth, our ribs are almost horizontal, which is why, when you watch a baby breathe, it is not their chest that moves but their abdomen. They are using the diaphragm, the sheet of muscle that separates the chest cavity from the abdominal cavity, like a set of bellows, to draw in air through the mouth and nose as the diaphragm contracts and blow it back out again when it relaxes. The ribs only start to take on the oblique angle we see in the adult at around two to three years of age. By this time, the pelvis has grown enough to allow the viscera in the abdomen to drop down and the little pot-bellied baby you've had for a couple of years appears, almost overnight, to turn into a scrawny little string bean who is now using their chest muscles to breathe.

Whether the ribs are horizontal or oblique allows the forensic anthropologist to narrow down a possible age for a child which can be confirmed by other parts of the skeleton, because our bones age as a collective, rarely in isolation. Each part of our anatomy talks to all the other parts so that they hum a similar tune and stay in harmony. It would be most unusual for one bone or organ to indicate that a person is in their fifties while another is suggesting they may be in their twenties. We don't have old chests and young legs. If that is what we are seeing, we are probably looking at two bodies.

So we use what one part is telling us to corroborate what others are also whispering. This system of constant checks and balances enables us to establish an age range and then decide whether the person is more likely to be at the top or the bottom of it. Age determination cannot be definitive. If any forensic anthropologist were to specify that an individual was twenty-three years of age, the police ought to be getting themselves another anthropologist, because that degree of precision is simply not possible.

Providing a range with an inbuilt margin for error also helps families to acknowledge that a body may be their missing loved one. If you give an age of twenty-three and their relative is twenty-five, it can be hard for them to accept that you may be two years out. A range of between twenty and thirty, with the suggestion that the person is likely to be somewhere in the middle, encompasses all the possibilities.

While ribs have some value in the determination of sex and age, they provide little if any information about ethnicity or height. It is when they have been subjected to trauma that they become particularly helpful to us in trying to establish a pattern or sequence of events before, during or after death.

The analysis of rib fractures in children has long been dogged by elements of controversy, especially in relation to the type of case often referred to as "shaken baby syndrome." All child deaths are emotive, and the dangers inherent in distinguishing between SIDS (sudden infant death syndrome) and intentional harm are starkly illustrated by the historical cases of Sally Clark, Trupti Patel and Angela Cannings, all of whom were convicted of killing their children and subsequently had their convictions quashed. Such miscarriages of justice, together with high-profile cases of shaken baby deaths, have, quite rightly, made paediatricians, pathologists and anthropologists cautious about how they interpret what they see when it comes to rib pattern fracturing.

Ribs tend to be a first port of call for investigators who suspect child abuse. However, a child's rib can fracture easily, and the explanation may well be innocent. Even multiple fractures can be due to one of several clinical conditions that can lead to brittle bones. In cases of SIDS, they can also be caused by attempts at resuscitation. It is vital that the fractures are viewed in the light of the general health of the rest of the skeleton, and the circumstances surrounding the child's life and death, before anyone jumps to the wrong conclusion.

The rationale for treating rib injuries as suspicious is that fractures may result from a child being grabbed by the chest and shaken violently, at the point where the hands of the person doing the shaking come into contact with the chest wall on either side. Breaks will generally heal within a few months in small children, often without leaving much evidence that they have ever happened. But where there is repeated child abuse, an X-ray can reveal fractures at different stages of healing: some from the past, which may be barely visible, others from a few months before, clearly still healing, and recent injuries showing little or no sign of callous formation.

A callous is new bone that grows around a break, acting a bit like a very big sticking plaster, to hold the two separated ends together and give them a chance to heal. Within a few hours, a haematoma, a large blood clot, will form around the site of fracture, producing a temporary soft-tissue callous (a bridge). This inflammatory response induces bone formation as new cells stream into the area to begin to repair the damage. Around seven to nine days after the trauma, the haematoma will have visibly been converted into a cartilage callous where bone can start to generate. Within three weeks, a bony callous has begun to form. Over time, several months, or years in some cases, the bone will be remodelled back into something close to its original shape.

When a child is being physically abused, there might well be other injuries in addition to fractures. In some of the worst cases, there can be little doubt that abuse has taken place, although it may be harder to prove who is responsible when there is more than one potential suspect. In one distressing case, there was no shortage of evidence on either count, but where forensic anthropology was able to help was in providing a detailed picture of what might have happened and when.

Harry was five years old when he died in hospital. His father had summoned the emergency services, telling them that he had found his son in bed, cold and unresponsive. Emergency staff were immediately suspicious when they registered the child's black eye and what looked like a deep bite mark on his cheek. On removing the child's clothes to administer CPR, their fears were confirmed. His body was covered in bruises and little circular marks that looked like cigarette burns. As they lifted Harry from his bed, they noticed a head wound. Police were of course quickly alerted. The full extent of Harry's prolonged misery was revealed at postmortem and following a radiological assessment of his injuries. It is terrifying reading the list, let alone trying to imagine what he lived through. These bare facts paint a picture of a truly horrendous short existence for a scared little boy.

The images from the PM and CT scans of his body were brought to my team in the hope that we could establish a chronological timeline for some of his injuries. Starting at the top, he had recent fractures to his skull which, we were informed, were the likely cause of death. Hair and blood found in the plaster of the bathroom wall suggested that his head had been banged repeatedly against it. On his face were four bite marks and a cut to his chin; his nose had been broken not long before he died, part of his earlobe was missing and he had two black eyes.

There were cuts and bruises to his legs and arms and numerous burns, some probably inflicted by lit cigarettes and others perhaps by an iron. His right arm had been broken recently, as had both bones of his right forearm. His left arm had been fractured in the past, and now showed quite extensive healing. Two breaks in his left forearm and fractures to the left thumb and fingers, and bones in his left foot, were new injuries.

The torso bore bruises and burns and he had been repeatedly punched in the abdomen and genitals. On the right-hand side, there were fractures to ribs 7 and 8—two to the latter, one older than the other, which suggested recurrent abuse. On the left, he had fractures of ribs 7, 8, 10 and 11, and again, two temporally distinct periods of fracture and repair could be seen on rib 11.

It is not within the forensic anthropologist's area of expertise to date soft-tissue injury. We work with hard tissues and, more specifically, bone. Focusing on the bones, we were able to determine from the images that the healed fracture to the left arm was probably up to a year old. Hospital records confirmed that Harry had attended to have a plaster cast applied. It was claimed that he had fallen in the playground. At least two of the rib fractures had been sustained about two to four months before he died. All the other skeletal injuries appeared to be recent, coinciding roughly with the time of his death. We could conclude that there had been at least three periods of repeated skeletal trauma. It was the fractures to his ribs that gave the clearest indication of the persistent nature of these events.

Harry had lived alone with his father since his mother departed the country, perhaps to escape the violence of her husband. But she had left her little boy behind to suffer this awful catalogue of abuse. The father pleaded a mental disorder, but the court did not accept this, and he was sentenced to life, and to serve a minimum of nineteen years before being considered for parole.

Child deaths are the hardest for every member of any medical or forensic team to deal with, but at the same time they instil a renewed sense of purpose and a drive to find the truth for the sake of justice.

The ribs run from the vertebral column at the back to the sternum at the front, where the bones are replaced by cartilage. We need the thoracic cage to retain as much flexibility as possible to assist us with breathing. The muscles that sit between the ribs, the intercostals, are responsible for raising one rib against the other. Because, in the adult, the ribs are not horizontal around the chest but at a curved angle, when they are raised, they work like bucket handles. This is why, when you take a deep breath, your chest swells not only at the front but also at the sides. This is caused by each intercostal muscle contracting and pulling up the rib below it, thereby altering internal thoracic pressure and drawing air through the nose and mouth into the lungs.

While the human normally has twelve pairs of ribs, this can vary. Some people have twenty-six, or even more in total. Cervical ribs (in the neck) can, if they grow too big, cause problems with circulation, pain and paraesthesia (loss of sensation) in the upper limbs. They can be surgically removed if they give too much trouble, with no ill effect on the patient.

These anomalies can naturally be a help in making identifications if their presence in skeletonized remains can be matched to a previous X-ray as their incidence is not frequent. And anyone with extra ribs that brought them discomfort is likely to have been referred at some point to a hospital.

Ribs in the lumbar region are more unusual, and arguably of limited value for identification purposes as they tend to be very small, almost vestigial, and generally produce no symptoms. Many people who have them are blissfully unaware of their presence.

Additional ribs may create some initial confusion when remains are discovered, especially if they are fully skeletonized. You might be down on your hands and knees in a muddy field, the horizontal rain of Scotland freezing your ears off, when you find an extra little bit of bone. You have to pay careful attention to where you found it, of course, but a lot of rain can move small pieces of bone around so the ribs aren't necessarily going to be located where you expect them to be. They might also have been disturbed by animals, of course. Since the calorie-rich viscera are a magnet to many scavenging creatures, parts of the torso may be dragged away from the site of the body for consumption and ribs can often show quite extensive damage caused by chewing or gnawing.

Sometimes it may be only ribs that we find, and then it can be tricky to establish whether they are actually human, especially if they are fragmented. Because in anatomy, form follows function, if a part of the body is doing the same job in one animal as it is in another, it may look very similar in different species of a similar size. This is particularly true of the ribs of humans and pigs. If you think about how often a police search includes a dump or landfill site, and the quantities of spare ribs sold in our restaurants and takeaways, it will give you some idea of how often forensic anthropologists are called upon to distinguish between them.

Readers of *All That Remains* may remember the mention made there of a character who claimed to have dissolved his mother-in-law in a mixture of caustic soda and vinegar. The victim, Zaina, was a fifty-six-year-old mother of six who took her youngest daughter to school one morning and was never seen again. Police investigations in her home revealed her blood in the bedroom, on the landing and in her bathroom. A crucial piece of evidence was a palm print, in her blood, found at the top of the stairs, which belonged to her son-in-law. When questioned, he produced many bizarre explanations for her disappearance and for the blood, including a tale involving her being kidnapped by masked men and held for ransom.

Eventually he admitted that Zaina was dead, alleging that they had always got on well but that one day she had made sexual advances towards him. He was, he said, utterly repulsed and had pushed her away, more forcibly than he had intended. She had fallen back and hit her head on the headboard of the bed; blood appeared at her nose, she didn't move and it was clear that she was dead. Panicking, he dragged her body across the landing into the bathroom and deposited it in the bath while he considered what to do. Fearing that he would not be believed, he decided that he had to get rid of her body. It was when he was questioned under caution that he said he had gone out, leaving her in the bath, and purchased a quantity of caustic soda and vinegar, which he poured over Zaina on his return. He told the police that her body had just dissolved and he had flushed her down the plughole.

It was at this point that the police came to me to ask if it was possible to dissolve a body in the way that the accused was describing. It was time to call a halt to his string of fantasies. First of all, domestic caustic soda is not sufficiently strong to liquefy a body, certainly not within the time frame given—a matter of hours. And combining it with vinegar would have neutralized the caustic soda. Vinegar is an acid, caustic soda is alkaline, and when you put them together you get something called sodium acetate and water. It's not a nice chemical, and there may have been some initial "burning' on the surface of the skin, but little more effect than that.

He had to come up with another story. Evidently, the son-in-law was no chemist, but he had relevant experience in another sphere. He worked part-time as a butcher in a pie factory and also served part-time in a kebab shop. You can see where the police thinking was starting to go. Using something like a cleaver and his not inconsiderable butchery skills, he had dismembered Zaina's body in the bath of her own home, wrapped the body parts in plastic bags and stored them behind the bar in the kebab shop. We know this because her blood was found there.

That night, he and his brother cut up the body parts into smaller pieces. They said they had then toured the city dropping them in bins outside other takeaway shops, where they would be picked up by the bin lorries and sent to a landfill site. As you can imagine, this prompted a full-scale food alert. But although all the waste destined for the landfill site, and the site itself, and the meat on sale from the takeaway and the pie factory, was checked, no evidence was found to confirm that Zaina had met this particular fate. It is perhaps not surprising that the kebab shop closed down shortly afterwards—although it reopened later under new ownership and, as far as I know, is still trading as a takeaway.

Zaina's son-in-law was given a life sentence and his brother a seven-year stretch for assisting in the disposal of her remains. Her relatives believe that the motive for her murder was money. Zaina had a nice house and cash in the bank, and the son-in-law wanted both. The pain suffered by a family in such circumstances—having to bear not only the devastating loss of a person they loved, but also the knowledge that their life has been violently taken away by a close relation, and the distress of never having the body returned to be properly laid to rest—must be unfathomable.

No body has ever been found, despite an extensive search that entailed identifying every spare rib recovered from the landfill site. Was it human or was it animal? Cases like this illustrate why it is so important for a forensic anthropologist to be as confident identifying fragments of rib from a pig, sheep, goat or other animal as they are pieces of human bone, and in their ability to differentiate between them. Students usually find ribs very boring and hate the hours we spend teaching how to tell them all apart, but we know that such a skill may well be critical to an investigation, especially one that involves dismemberment.

We also drum it into our students that they must be able to seriate the ribs, which means knowing how to distinguish right from left and even, in the absence of a full set, which region of the chest they are likely to have come from: top, middle or bottom. Can you tell, for example, whether what you have is part of the right fifth rib or the left fourth rib, even when all you have is a fragment? This is by no means easy to establish but it can matter.

I once gave evidence at trial in relation to the dismemberment of the body of a baby boy whose skeleton had been found under a concrete floor. The defence pushed hard on how certain I could be that the knife used had entered between the fifth and sixth ribs. Was this vital to the case? Probably not, but deliberately unsettling an expert witness in order to introduce just enough doubt in the minds of the jury as to the validity of their testimony ("What do you mean, you can't be sure? What kind of an expert are you?") is a courtroom tactic frequently employed by defence lawyers.

Telling right from left is relatively straightforward, as long as you have the segment of the rib next to the vertebral column at the back. Here, where the shaft of the bone turns a corner, a furrow starts to appear along the bottom border of the rib. This is the subcostal groove, which houses blood vessels and nerves that run all the way along the length of the lower border of the rib from the back to the front.

If you are a meat-eater (and it doesn't put you off your dinner), you can see this for yourself the next time you eat spare ribs. Look at the area of muscle closest to the bone. Providing it is a lower border of the bone and not an upper one, the holes that convey the blood vessels, and a white, solid little rod, the intercostal nerve, should be visible. Because the subcostal groove is always on the bottom of the bone, and the outer surface is convex and the inner one concave, we can tell which way up it goes and front from back. This enables us to say reliably whether the rib is from the right or the left. It sounds logical, and it is, but it has to be taught.

Our first task, then, when seriating ribs, is to quickly sort rights from lefts. If all the ribs are present and correct, we expect to have twelve on one side and twelve on the other, but of course things aren't always that simple. There may be extra ribs or some missing because of damage or scavenging. Now what we have to decide for each rib is whether it comes from the apex of the chest, the upper middle region, the lower middle region or the base of the chest.

The first two ribs don't look like any of the others. The really tight angle they have around the top of the lungs gives them a distinctive "comma" shape which makes them easy to identify. The next four (ribs 3–6) are what are called "true ribs," or vertebrosternal ribs, as they each have their own separate costal cartilage that attaches to the sternum at the front and their shape reflects this upper-middle function.

The lower-middle ribs (7–10) are known as false, or vertebrochondral, ribs because their anterior ends do not extend all the way round to the sternum, terminating instead in a common costal margin, which you can see quite clearly in individuals who do not have a lot of covering fat. The final two (11 and 12) are called "floating' ribs as they don't attach at the front to either the costal margin or the sternum and simply terminate in the muscle of the abdominal wall. As a result, they are somewhat vestigial and much smaller.

In some extreme types of cosmetic surgery, people may choose to have their lower ribs shortened or even removed altogether. The fashion for an exaggerated hour-glass figure, which the Victorians achieved with very unforgiving corsetry, can today be surgically acquired. More meaningfully, the lower floating ribs can be taken out and used as an autograft to repair fractures elsewhere in the patient's body, such as the face or mandible. A very dear friend of mine, who was a paramedic with the Marines, was shot on a tour of duty in Northern Ireland while trying to recover a wounded soldier. One of his ribs was very successfully pressed into service to rebuild his shattered jaw. It is useful to know which bits of your body are not essential to you and can be used as spare parts if required.

Sometimes it isn't just the ribs that need to be identified, but matter from elsewhere in the body that becomes associated with them. I once found, nestled on the inner surface of the back of the right ribs of an elderly lady, a mass of tiny stones that would have been inside her abdominal cavity. These were gallstones, formed as a result of her diet containing too much cholesterol and her liver producing insufficient bile salts to dissolve them. Gallstones can accumulate in the gall bladder, which is essentially a little storage pouch, or move to block either the bile duct or the sphincter at the opening connecting the gall bladder to the small intestine. This lady had some very large stones, about the size of walnuts, and many smaller ones like sweetcorn kernels, with flat faces and sharp angles where they fitted into one another like a jigsaw. Stones also occur in other parts of the body connected to the urinary system, including the kidneys, ureters and bladder. So we need to be mindful of the "stones within the bones."

Seriation and identification of ribs is relatively uncomplicated when they are intact and adult. Seriating infant ribs is another matter, and requires specialist knowledge and experience.

In 1999 I was called upon by the Foreign Office to fly out to Grenada in the West Indies to help with a situation of some political sensitivity.

Grenada had gained independence from the UK in 1974, with Sir Eric Gairy becoming the country's first prime minister. Five years later, while he was away at a UN summit, control of the country was seized in a bloodless coup by a revolutionary group called the New Jewel Movement, or NJM ("Jewel" was an acronym for Joint Endeavour for Welfare, Education and Liberation). Their leader, Maurice Bishop, viewed as a heroic "man of the people" by many Grenadians, dissolved Parliament and appointed himself head of the ruling PRG (People's Revolutionary Government).

The revolution was welcomed by most of the population and Bishop set about implementing a range of measures to improve the lives of the islanders, including free education and healthcare, better public transport and new infrastructure projects. But it was not long before cracks started to appear within the PRG and in 1983 he was deposed and placed under house arrest by members of his own party loyal to his second-in-command. The country descended into chaos.

A crowd of several thousand of his followers freed Bishop and marched with him to the army HQ. A military force was dispatched from another fort to quash the protest and eight people, including Bishop, three Cabinet ministers, among them his girlfriend Jacqueline Creft (who had been minister for education), and two union leaders, were taken away. They were said to have been lined up against a wall and executed. It is not known what happened to their bodies but there were many rumours, including claims that they had been placed in a pit, had petrol poured over them and were set alight, with grenades being thrown in to blast them into unrecognizable pieces.

US president Ronald Reagan ordered an invasion of Grenada, citing concerns for the safety of several hundred American medical students who were resident on the island. Although Margaret Thatcher, the UK prime minister, was unimpressed not to have been consulted about the invasion of a former British colony, the British government remained publicly supportive of the US decision.

Operation Urgent Fury, a four-day land, air and sea offensive involving over 8,000 US forces, swiftly restored peace, but Bishop remained a national martyr without a fitting grave. Several attempts were made to find his remains, and a US military investigation had also apparently been unsuccessful.

The phone call from the Foreign Office came after a gravedigger unearthed a US Marines body bag containing human bones in what should have been an empty plot in a cemetery in the Grenadian capital, St George's. Rumour spread like wildfire among the locals that these could be the remains of the PRG martyrs, perhaps even of Maurice Bishop himself, and there were concerns about potential unrest. A combined task force from the US military and the FBI was preparing to fly out to examine the remains and the Grenadian government contacted the UK government to request a small and impartial observer team to join the investigation.

Our team was indeed small. It consisted of me, the forensic anthropologist, and Dr Ian Hill, the forensic pathologist. The contrast between the Brits and the Americans couldn't have been more striking. The large US contingent arrived in force, all corporate jackboots and polo shirts, baseball caps and jackets bristling with logos. They boasted the latest gear, packed into shiny, metallic matching luggage, and an air of distance and superiority that was palpably chilly.

Ian looked like the man from Del Monte with his very English sun hat, short-sleeved checked shirt, cream-coloured blazer and beige slacks. I, as usual, looked like someone's mother (which, to be fair, I was). We were clearly instantly assessed as "no immediate threat," supernumeraries to be tolerated but largely ignored. Always a dangerous assumption, as any fan of *Columbo* knows.

Ian, an RAF man, spent the entire flight out to Grenada commenting on every creak and groan of the aircraft, recalling every airline crash he had ever attended and advising me on what I should do in the event of an emergency landing. If I was calm on take-off, I was a nervous wreck by the time we touched down. I was glad to arrive at the lovely hotel where we were billeted. It always feels incongruous, on a forensic mission, to find yourself on a tropical island surrounded by swimming pools, cocktail bars and open-air restaurants, but we soon adapted.

Our first meeting with the US forensic team prompted them to reassess their first impressions of us, if only to add the adjective "annoying" to their appraisal. We asked for a copy of the previous US investigation's report. We were told in no uncertain terms that they were presently unable to locate a copy, but "ma'am" should be reassured that as soon as they did, we would be supplied with one. There was no attempt even to pretend that this rhetoric didn't smell like something coming straight out of the back end of a large male bovine.

Ian and I decided on a strategy of attrition. We just kept on asking, every single day, if they had yet laid their hands on a copy of the report. Every day we got the same stock reply, delivered with patience and courtesy. Some days, just to break the monotony, we would ask twice, or take it in turns, with Ian asking one day and me the next. Or we would both ask, as if we didn't communicate with each other. Sometimes you just have to make your own fun.

We sought out the gravedigger to talk to him about what he had found, where he had found it and why he thought everyone was so twitchy. A lovely man, he was happy to take us to the cemetery and show us the hole he had refilled once he realized what he had stumbled upon. He passed on local opinion on what was buried there and imparted all sorts of useful intel—much of it gossip, but some of it relevant detail that had not been reported in the preliminary briefing meetings. He told us it was common knowledge that Jacqueline Creft had been pregnant with Bishop's child at the time of their execution. Since this was a critical nugget our US colleagues had chosen not to share, we hadn't been factoring in the possibility of recovering fetal remains. Which just goes to show how important it is to speak to everyone, however peripheral their involvement may seem.

We started the official excavation very early in the morning to try to avoid the searing midday heat. Not far below the surface we encountered the anticipated US Marines canvas body bag. It was in poor condition but we were able to lift it virtually intact using the rolling technique perfected by nurses for changing sheets under a bed-bound patient. We employ this method regularly when we need to transfer remains into body bags with minimal disruption.

The US team began to pack up as soon as the bag had been extracted. But that is not our protocol. We would always ensure that a burial site is searched fully, both below the recovered remains and to the sides. In this case the body bag had been breached, and bones can move within the soil in all directions as a result of faunal activity and the effects of water courses in the soil. So we would never have assumed that the bag and its contents were the only items in the hole.

As I trowelled away some soil, I started to uncover some additional bones. That is, of course, only to be expected in a cemetery that has been in use for many years. These were small, juvenile human ribs. They would, naturally, have to be investigated, but it was clear to me immediately that they were not fetal, and therefore there was no possibility that they could be associated with Jacqueline Creft's rumoured pregnancy.

However, I was in a mischievous mood. I looked up from the depths of my dusty, dirty hole in the ground at the gleaming jack-boots of my US counterpart and asked him sweetly, "Tell me, are we expecting to find a child?" The blood drained from his face and, without a word, he turned and ran to an isolated corner of the cemetery, gesticulating wildly and gabbling into his mobile phone. I won't deny that this small chalking up of points on the UK scoreboard was childish, but I couldn't help allowing myself á self-congratulatory little smile.

When he returned, I asked him if he was comfortable seriating juvenile ribs. If so, I would just hand them up to him. I thought he was going to faint. It was clear that he had little experience in dealing with juvenile remains. I did not mention, as I passed the bones upwards, that they were not fetal, because of course I wasn't supposed to know that the possibility of finding fetal bones was even on our radar. He asked me how old I thought the child might be.

"Young," I said. Perhaps I enjoyed this just a little too much, but the Americans had been so incredibly standoffish to us that I felt I deserved some tiny payback.

In the mortuary, I prolonged his discomfort through what had now become a role reversal by giving him the partial juvenile skeleton to lay out. He spent the next four hours setting up his "equipment," and on his mobile phone because, apparently, he was having some technical difficulties. Of course, you don't need equipment to lay out a juvenile skeleton, just experience. I left him to it until about an hour before the end of the day, at which point I finally put him out of his misery and laid out the bones myself, from head to toe, in about fifteen minutes. When I announced that this was a child of about two years of age, he did actually offer me a weak smile. I felt I had made my point, and he knew it. Thinking that maybe I had scaled a barrier, I pressed my advantage by asking again for the report of the previous investigation. But apparently it was still missing in action.

Ian and I got on so well with the gravedigger that he invited us to a party in his garden that evening, and of course we were always going to accept. He issued a similar invitation to our US colleagues, who declined politely but firmly. It was their loss.

When we arrived at the gravedigger's house, our generous host took us out into the garden, where an enormous pot was bubbling over an open fire. The pot, lined with bread, was piled high with chicken and vegetables cooking in the middle so that the bread soaked up all the juices. The scent was divine. Then he broke open a demijohn of homemade demerara-spiced rum and the evening became even more convivial. We knew we would suffer for it the next day, but it was worth it.

As the night wore on, we all sat under the moon around the fire, mellowed by rum and great food, and shot the breeze. I got talking to a charming man who turned out to be the professor of anatomy at the island's private university. Of course, we had a lot in common and spent a good deal of the rest of the evening in discussion over matters anatomical, educational and anthropological. While exercising the caution demanded by the political sensitivities in talking about the reason for my presence in Grenada (although, as it turned out, I needn't have bothered, as apparently it was hot gossip all over the island), I commented on how funny, and infuriating, Ian and I were finding the intransigence of our US colleagues in the matter of the elusive report from their previous investigation. Then, beauty of beauties, came words I had never expected to hear.

"I was around at the time and I have a copy of that report. Would you like me to photocopy it for you tomorrow?"

It is astonishing what magic can be woven from the ingredients of a pair of anatomists, a gravedigger's party and homemade demerara-spiced rum. The child's ribs were going to be the least of my colleagues' concerns in the morning.

The next day, at the university, our hangovers in full throb, Ian and I made our way to the professor's office where the copy of the US report was waiting for us. In fact it contained very little that we didn't already know, other than a bit more background on the theories suggested at the time of the first, unsuccessful, mission to identify the location of Bishop's remains, and, of course, the information that Creft had been pregnant, which we were now aware of anyway.

Ian and I returned to the mortuary, where our US colleagues seemed to be very busy packing up, posed our ritual morning question—"Have you managed to find the report yet?"—and received the ritual response: "Sorry, ma'am, not yet. Still looking." I produced the report from my bag and asked them if perhaps they would like to make a copy of ours, as we would be really happy to assist the USA in completing their records. Well, they were like rats up a drainpipe. The entire team leaped from their chairs, raced out of the room, phones clamped to their ears, chattering nineteen to the dozen. The strange thing is, the Americans never did take me up on the offer of a copy of our report. Maybe they managed to find their own after all.

In the end, there was nothing for anyone to worry about. None of the remains found were those of Bishop, his girlfriend or any of the Cabinet members.

When the Americans invaded Grenada in 1983, they had launched an airstrike on army headquarters, missed their target and instead hit an asylum hospital building nearby. While the child's ribs were likely to have come from a previous burial in the cemetery, the other bones in the grave, including the ones in the US Marines body bag, were those of several unfortunate patients from the hospital who had simply been in the wrong place at the wrong time. We were able to confirm this from the fragmented nature of the body parts, the mix of sexes and ages and a piece of pyjama waistband still, poignantly, bearing a label with the hospital's name stitched into it.

It is amazing how much mischief can be had with a set of ribs and a few sheets of paper, especially when people forget to treat each other with the respect we all deserve and when they choose to be divisive rather than collaborative. It was so unnecessary when all any of us was trying to do was to get to the truth.

# 5

# The Throat
## *Hyoid and Larynx*

*"The human voice is the organ of the soul"*
Henry Wadsworth Longfellow
Poet, 1807–88

If one bone might be said to be the favourite of crime writers, it is the hyoid bone. Because of its susceptibility to fracture, many is the death by strangulation attributed to such an injury in novels.

The hyoid bone sits in the upper region of the neck, below the jaw, in front of the third cervical vertebra. If you place your fingers on either side of your neck in this area and squeeze (not too hard), you may feel resistance on both sides and a rather unpleasant pain in the neck. The bones underneath your fingers are the tips of the greater horns (or wings) of the hyoid bone, which acts as a sort of staging post for the attachment of muscles from the lower surface of the jaw towards the top of the bone, and for the muscles that pass from the lower end of the bone down to the sternum and other structures around the larynx or voicebox below.

In a child, the hyoid is made up of five separate pieces of bone: a piece in the middle (the body), two lesser horns to either side and, below these, the two greater horns. It is roughly C-shaped, with the open part of the C surrounding the windpipe, facing towards the back. The lesser horns will fuse to the body early on in life but the greater horns may not start to do so until the fourth or even fifth decades.

Pressure placed at the sides of the neck between the thumb and fingers of an assailant can break the rather flimsy greater horns, but the truth is that not all strangulations result in a fractured hyoid. It is estimated that around two-thirds do not. Indeed, such fractures are not common in young people and very rare in children. And even if a dead person is found to have a broken hyoid, it doesn't necessarily mean they have been strangled, as this fracture could have occurred during their life.

Jenny's sad story is a case in point. She had had a troubled childhood. Both of her parents had died and she had been placed in foster care separately from her brother. She went on to have three children of her own at a young age and, following the break-up of her marriage, her already haphazard life began to descend into chaos. Known to be a heavy drinker and to be involved in substance abuse, she frequently disappeared for weeks at a time, sleeping on friends' sofas, dossing in abandoned properties or occasionally booking into a hostel if she had a little money. She was thirty-seven years old when she was reported missing by somebody who realized they had not seen her for quite a long time.

Her last-known abode had been an empty house on the outskirts of a northern city. Neighbours of this property frequently complained to the council about vagrants using it as a squat and fly-tippers dumping rubbish there. Nineteen months after Jenny had last been seen, a public health order was placed on the house and a company was contracted to clear it out so that it could be done up and sold. In the back yard, under six feet of rubbish bags and general detritus, they were shocked to encounter human skeletal remains, curled up in the fetal position. Close to the body were an aerosol can and a plastic bag.

The bones were sampled for DNA and they were confirmed as being Jenny's. It seemed she had died in the back yard of the house and that, as fly-tippers had thrown more and more rubbish over the wall, she had gradually been buried deeper and deeper under the accumulating mass of refuse.

Body recovery was undertaken by a forensic archaeologist to ensure that as much of the remains as possible could be retrieved.

The postmortem findings were inconclusive as regards cause of death, although several healed fractures were found in many bones of her body. She had clearly had a life of literal hard knocks but what could not be established was whether death was by misadventure or something more sinister.

The archaeologist had recovered the hyoid bone in four pieces and identified them all—a master class in professionalism. The lesser horns were fused to the body. The right greater horn was separate, and the left one was also detached and in two pieces. The question we had to try to answer when the police brought the bone to our lab was whether the fracture of the left greater horn had been caused antemortem (before death), perimortem (around death) or postmortem (after death).

A premortem injury would be likely to show some signs of healing, whereas a perimortem or postmortem fracture would not. A perimortem fracture is generally not a clean break because the bone is still wet. Imagine trying to snap a green twig. The broken ends will usually be ragged, with strands of stubborn, straggling wood hanging from them. When you snap a twig that is dead and dry, the break tends to be clean, as do those of bones fractured postmortem, after they have dried out.

Premortem injuries may be the result of violence or trauma that occurred some while before death. Postmortem fractures are generally caused by the treatment of the body after death or during its excavation. One will show healing and the other will not. Perimortem fracturing, however, speaks to us of a possible violent cause of death and may well prompt a homicide investigation. Trying to establish as accurately as possible when the wing of the hyoid had been broken could be the key to understanding something of the nature of Jenny's death.

With the naked eye, the two separated surfaces of the left greater wing of Jenny's hyoid looked pretty clean, but under the microscope we were able to see something very different. The bone had been fractured when it was wet, so when she was alive, and the rounded appearance of the strands showed that it had been trying, unsuccessfully, to heal itself. Jenny had lived with her fractured hyoid for quite some time after the event that had caused it, though probably only for months rather than years.

Multiple healed fractures may be an indication of domestic abuse or assault. They might also simply be evidence of a lifestyle of habitual intoxication resulting in regular falls. There was no definitive evidence of assault in Jenny's background but there were significant hospital records of admission to Accident and Emergency due to falls, especially in bad weather.

Our middle daughter worked for a time in the orthopaedic ward of a city hospital, where she would see many admissions for fractures sustained in falls while under the influence of drink or drugs, frequently when it was raining heavily, or when the first ice of winter settled on the pavements and roads. She recalls how traumatic it was for the nursing staff to try to deal with the complex needs of these patients while simultaneously staying vigilant about health and safety, because you had no idea what infections they might be carrying. Often staff had to treat injured people handcuffed to beds, with police escorts present, because of their violent outbursts as they went through withdrawal. It is a chaotic life not only for those, like Jenny, who are living it, but also for those who are attempting to help them.

However, it is quite difficult to fracture your hyoid in a fall, and so in Jenny's case, suspicions that she may have previously suffered an assault remained. It was not ultimately possible to determine her cause of death, but whatever had happened, it was unlikely that the broken hyoid had any crucial role to play in it.

The major parts of the voicebox below the hyoid bone are the thyroid and cricoid cartilages, which can begin to turn into bone with advancing age. This ossification can produce an incredibly delicate and strangely beautiful lace pattern of bone formation as we grow older.

The thyroid cartilage, better known as the Adam's apple—probably a reference to the ancient belief that a piece of the forbidden fruit got stuck in Adam's gullet—is usually more developed in males as a result of changes to the voicebox during puberty, when the cartilage increases in size and the voice drops. The vocal cords are attached to the back of the thyroid cartilage, and so the more prominent the

Adam's apple, the longer the cord and the deeper the timbre tends to be.

Laryngeal growth is not as pronounced in females, although it can be present in some women with a larger larynx. Generally, though, the prominence of the thyroid cartilage is so strongly associated with masculinity that it can be a cause for concern among those transitioning, especially from male to female. Often scarves, chokers or high-necked clothes are worn to disguise it. It is possible to have the cartilage "shaved" to reduce its size, and some choose that option.

Bone can start to be laid down in the thyroid cartilage as early as the third decade, but the timing is very variable, and there are no obvious sex differences in when ossification of the cartilage begins.

The cricoid cartilage lies below the thyroid cartilage, level with the sixth cervical vertebra. It is shaped like a signet ring, with a broader surface towards the back and the narrower band at the front. Below this is a series of cartilage rings that keep the trachea open so that we can breathe, and these, too, can turn into delicate little bony rings with age.

The hyoid, thyroid and cricoid bones, as well as ossified tracheal rings, are among the assorted funny little bits and bobs that may be presented to the forensic anthropologist by the human skeleton, and which we therefore need to be able to recognize.

The constrictions of this area of the neck can of course be a hazard during our lives should a foreign body enter the airways. I remember one case history I read about while researching for a textbook I was writing. The patient was admitted to A&E at Christmastime in extreme respiratory distress. He thought he had swallowed a turkey bone. During an oesophagoscopy, the foreign body was detected near the thyroid cartilage and extracted.

It turned out to be not a turkey bone but a piece of shell. When asked again to try to list exactly what he had eaten, the patient recalled that the turkey had been stuffed with oysters. So you never know what you might find when you go looking.

# PART III

# THE LIMBS
## Postcranial Appendicular Bones

# 6

# The Pectoral Girdle

*"Shoulder blades are where your wings were when you were an angel"*

David Almond

Writer

There are two "girdles" in the human body. The word "girdle" is more commonly associated with women's corsetry, but I gave up using such analogies with my students when references to my mother's Playtex girdle and Cross-Your-Heart bra drew nothing but blank looks. Evidently I was showing my age.

Our upper bony girdle is the pectoral or shoulder girdle, which connects the bones of the arm (humeri) to the trunk and comprises paired clavicles (the collar bones) at the front and scapulae (shoulder blades) at the back. The lower one, the pelvic girdle, consists of the two hip bones, which form a junction between the sacrum at the back and the femora (thigh bones) of the lower limbs at the sides.

It is interesting that the pectoral girdle should contain both the bone that is the least likely, of all the bones in the body, to fracture—the scapula—and the one most prone to being broken: the clavicle. While all primates possess a collar bone, it is rudimentary in many mammals and absent altogether in the ungulates, which include a variety of animals from horses to pigs, and even the hippopotamus. Cats, for example, have very rudimentary clavicles, which is why they can squeeze through spaces that appear to be much too narrow to accommodate them.

In humans, the clavicle, as well as being a convenient location for muscle attachment, serves as a strut to keep our arms out to the sides of our body. In most quadrupedal animals, the forelimbs, positioned underneath the body, are used solely for locomotion, and as no dual function for the clavicle is required it does not need to be very large. Yet amazingly, the human clavicle isn't really essential. We can have it taken out as long as the muscles can be stitched to each other. In the past, some jockeys used to have their clavicles surgically removed as a preventative measure. Since it was the bone most often broken in falls from their horses, there was a school of thought that maintained it was better to do without it than to risk the perils of a fracture.

And there is no question that a broken clavicle can be life-threatening. The bone is shaped like an elongated "S' and a fracture will occur at the weakest point, the major bend in the lateral third. Unfortunately, that lies directly over the subclavian artery and vein, which are very large, and makes them susceptible to being ruptured or pierced by sharp shards of broken bone.

Sir Robert Peel, who served twice as Britain's prime minister between 1834 and 1846 and is regarded as the father of modern policing (hence the antiquated nickname "peeler," and the more enduring "bobby," for a police officer), met his end as a result of a fractured collar bone. He had acquired a new horse, a hunter, which had a bit of a reputation as a kicker. Sir Robert and the horse were still getting used to one another when, on his way up Constitution Hill, close to Buckingham Palace, he encountered two ladies of his acquaintance and their groom, who was on a rather skittish horse. Peel's horse was spooked and threw him off, and then, unfortunately, stumbled and fell on top of him. He suffered several broken ribs and a broken left clavicle, which ruptured the subclavian blood vessels underneath, and he bled to death.

As it took him nearly three days to die, and given the range of additional crush injuries he could have sustained, I suspect that the precise cause of death is more likely to have been due to other complications, but the story of the fractured clavicle has persisted through the years and it is still cited today as being responsible for his demise.

The clavicle is the first bone in the human body to start to form and it does so in the fifth week of intrauterine life, possibly before Mum has even realized she is pregnant. It is a very precocious bone, adopting its adult "S" shape very early on, by the end of the second month of pregnancy, and it grows at a very regular rate of about 1 mm a week from that point onwards. By the time the baby is born, it is about 44 mm long and very recognizable, and therefore particularly useful as an indicator of age in fetal and newborn remains.

While the prospect of a new life coming into a family is for most people something to be celebrated, sadly, not all babies are welcome and it is not uncommon for the remains of fetuses or newborn babies to be found concealed in unexpected places. Often it is when floorboards are lifted or old bath panels prised off, when chimneys are opened or swept, when lofts are being insulated or when old suitcases are discovered at the back of long-forgotten cupboards. Unwanted pregnancies can be covered up and the tiny body of a newly delivered baby, whether born alive or dead, easily hidden in the belief that no evidence of its existence need ever come to light.

But frequently it does, sometimes years after the event. Many of the cases we are asked to examine date back seventy years or more, to a different time. Terminations were illegal and unsafe, but that did not deter women from seeking them, or hinder the back-street industry that met the demand. Women were often driven to this course of action by poverty, and the economic impossibility of feeding another mouth, or by the shame and stigma attached to illegitimacy.

The discovery of such remains raises many questions. The first that spring to mind, of course, are when did the baby die and who was its mother? But often the most relevant questions from the legal point of view concern how and at what age the infant died. The central issue here is whether the child was born alive, and if so, whether it then died of natural causes, perhaps as a result of the absence of medical intervention, or was killed by somebody. If it died prematurely, was this perhaps due to a termination, or was it stillborn?

A stillbirth is defined as a baby born without signs of life after twenty-four weeks of gestation, whereas a child who dies before that stage is viewed as a miscarriage or late fetal loss. The twenty-four-week milestone is important forensically, as it is the current legal limit for abortion and thus deemed to be the age beyond which the fetus has a chance of survival, provided a high level of medical care is available. In other words, it is when the fetus is said to become technically "viable."

The fetal clavicle can provide reliable evidence in establishing this legal distinction. At twenty-four weeks, the bone is around 27 mm long—half the length of an adult thumb—and can be measured accurately. In a living baby inside its mother, this is done using ultrasound. The image can be quite awkward to interpret so the skilled advice of a radiographer is generally required. In a baby no longer in the uterus, an X-ray or CT scan of the bone can be taken, or, if a postmortem examination is to be performed, the clavicle can, of course, be removed and measured directly.

Finding the remains of fetuses or newborn babies in your house can be quite traumatic for the homeowners, as one couple who bought an old stone croft cottage in a remote part of the Scottish isles could attest. While undertaking some extensive renovations, they pulled up the kitchen floorboards to put in a damp course and new pipework. As they peered down at the earthen foundations below, they could make out what they thought might be bones on the surface of the soil. The island had a rich heritage of ancient burials and artefacts, so they called in some archaeologists working at a site nearby to have a look. The bones were very small. Some were those of animals, but, unfortunately, not all of them. The police were called.

As the local force had no scene-of-crime officers and it would have taken a couple of days to bring in SOCOs from the mainland, they decided to enlist the help of the archaeologists. The dry bones were lifted and transported by air, in two little cardboard boxes, to the nearest mortuary, over 150 miles away in Inverness. I was asked to examine the remains and give an opinion on their age at death, how long ago death may have occurred and to provide any other pointers that might assist their inquiries. The quality of the photographs taken at the location was extremely poor, to the extent that I had to ask what I was supposed to be looking at and where. That was alarm bell number one. Alarm bell number two rang when I asked who had lifted the bones and was told it was "OK because they were archaeologists, and they only picked up the human remains and threw away the animal bones."

Anyone with forensic training knows that you never throw anything away and should always use experts with the relevant expertise. Even so, this might in the circumstances have been "OK" if I'd been able to depend on the accuracy of the archaeologists' assessment of the origin of the bones. If they'd had the necessary experience to competently identify human remains, there should have been no animal bones in the boxes. But what I found were animal bones mixed in with human fetal bones. That told me I could have no confidence in the ability of the archaeologists to distinguish between them.

The disposal site would need to be searched again. When it was, I don't believe anything further was found, although I don't think a forensic anthropologist was present even then. Real-life investigations are never like the ones we see on television, and there was to be no eureka moment. I gave the senior investigating officer (SIO) a bit of a hard time about the quality of the photographs and comprehensiveness of the recovery and told him I hoped this wasn't going to turn out to be a homicide, because the evidence was patchy. He was a longstanding friend of mine, and he took it on the chin. But I suspect the experience nudged that police force into sharpening up their search and recovery procedures from then on.

The animal bones in the boxes were small, and from vermin, mainly mice and rats, and it therefore seemed likely that the human material had been a food source for them over time. Indeed, animal gnawing marks were visible on some of the human bones. A new-born baby has over three hundred bones and I was looking at only about 2 per cent of that total. Moreover, it was clear that these had come from more than one baby. Among them were three clavicles: two left and one right, and the right clavicle was not the same size as either of the left clavicles, so obviously not part of a pair.

In short, we had the remains of at least three different babies. Provided all three had been buried intact, I would have expected a minimum of nine hundred bones to be found. The rest had probably been lost over time: either consumed by animals, washed away (the area was very damp) or simply disintegrated by the local acidic, peaty soil. But of course, it may have been that some of them were just not picked up by the archaeologists because they had not recognized them as being fetal remains. It was also possible that the babies had not been discarded intact, and I would certainly be checking for dismemberment marks.

There was no evidence of trauma on the bones, so a cause of death could not be readily determined and dismemberment was unlikely. You cannot tell the sex of a baby from its bones, but you are able to pinpoint its age with accuracy, especially if you have the clavicles. Two of the babies, those for whom we had the left clavicles, were full term, about forty weeks old when they died. The third, the owner of the smaller right clavicle, was much younger, about thirty weeks— still viable according to the legal definition in force today, although potentially not capable of survival if the remains were of historic origin. Bone was sampled for DNA but none could be extracted, perhaps because of preservation conditions or its age.

We believed the bodies were probably historical. Radiocarbon dating could have confirmed this, but I am always reluctant to send baby bones away for testing unless absolutely necessary, especially when so few have been recovered. The analysis requires so much bone to be destroyed that there is a risk there may be nothing left to bury after our questions have been answered. So I asked the police to do some background research and set aside chemical testing as a last resort.

The story they uncovered came largely from local hearsay but it fitted the evidence and would eventually satisfy the procurator fiscal. It dated back to the days after the First World War, when this remote island community led an isolated existence, with no telephones, no electricity, no running water and no public transport. Life was harsh and most families subsisted on the meagre pickings they could scratch from the land or the sea. The houses were small, cold, damp and dark, with thick stone walls, thatched roofs, tiny windows and floors laid directly on top of the soil.

Violet, who was unmarried, lived alone in a typical little stone "but and ben" only 100 yards from the cottage where the bones were discovered. She was viewed by the local gossips of the time as a woman of dubious morality, and described variously as a floozie, trollop and Jezebel, or, in Gaelic, a *siùrsach* or *strìopach*.

It was said that, to make ends meet, Violet would sell her favours to the servicemen stationed at a nearby naval base and to well-off local businessmen. She was periodically seen wearing suspiciously baggy clothes and there were times when she moved in for a while with her domineering mother, Tamina, who lived in the nearby cottage, re-emerging later and returning to her normal life. It is what happened during these spells at the cottage that may have been relevant to the discovery of the remains found under the floorboards.

In an era of less than perfect contraception, unwanted pregnancies were an occupational hazard of Violet's alleged trade. It was claimed locally that she had given birth to anything up to eleven children in total, although gossip is always prone to exaggeration. Whatever the truth, by the time she died in the 1950s, she had only one surviving child, a son. His birth, it was remembered, had been a breech delivery and had required the attendance of a local doctor. This may have been what saved his life.

It was said that, with no means of terminating a pregnancy, Violet carried these babies to full term and moved in with her mother when the time came for her to give birth. Perhaps her relatives chose to look the other way when she became pregnant, hoping that she might miscarry. Perhaps they even benefited from the money she earned. Whatever the case, in those days illegitimacy was a sin in the eyes of the kirk and a stain on the reputation of the whole family—and Violet's grandfather was a lay church minister. While a blind eye might be turned to a discreet pregnancy, a bastard child would not be tolerated. And yet it seems that infanticide was. The condemnation of the Church was feared far more than the long arm of the law.

Local lore had it that, as soon as a baby was born, Tamina would take it away and drown it in a rusty old bucket normally used for carrying fish. The body would then be thrown under the floorboards of the cottage where, over time, it would decay until there was nothing left but bones.

Violet's son, who was no longer alive when the remains were found, maintained that, on her deathbed, his mother admitted to having given birth to five babies, and had said that his grandmother had drowned the other four. She told him he owed his life to the doctor's presence at his birth, which meant his arrival had to be acknowledged and questions would be asked if he suddenly disappeared. Otherwise, Tamina might have drowned him, too.

He never really knew his formidable grandmother. Violet was so scared of Tamina that she kept her little boy out of sight until he was old enough to go to school, when there could be no concealing his healthy existence.

There is, of course, no evidence for any of this, and most of it is probably salacious scuttlebutt. And before any of us rushes to condemn Tamina as a cold, evil serial killer, we need to consider the attitudes of the times. The deeds of the past do not always sit comfortably with our modern morals. Maybe Violet sought the help of her mother; maybe they worked together to maintain the family's meagre income, ignoring the gossips and disposing of the social embarrassment.

Illegitimacy and infanticide were sufficiently prevalent that in 1809 the law in Scotland had been amended to reduce the sentence for the offence of concealing a pregnancy and failing to call for assistance at the birth. Since the seventeenth century, this crime had been treated as murder, but it now carried the far less severe penalty of two years in prison. And if charged, women could claim stillbirth in the hope of being granted leniency.

If, as Violet's son had said, there had been four other babies, I could confirm the remains of only three under the floorboards. But it was possible that some of the bones could have belonged to a fourth baby or, of course, that its remains had been consumed in their entirety by scavengers or had been deposited elsewhere. Forensic anthropologists record an MNI—minimum number of individuals—which does not mean that more may not be represented. The MNI is calculated by establishing where there is duplication of the same bone, or bones of different sizes that may indicate different stages of maturity. We knew we had three clavicles that did not belong together, but we had no way of knowing whether all of the other bones were from the same three babies.

However many there were, some eighty years after the infants had lost their lives, allegedly at the hands of a murderous grandmother, they were finally laid to rest beside the woman who was believed to be their mother. They had no names, but the meagre pile of bones was placed in its own tiny casket. We could not prove that they even belonged to the same family, but anecdotal evidence and circumstance seemed to support the probability. The procurator fiscal was content and surely, after years lying beneath that cottage like unwanted waste, they deserved to be buried with respect.

At around the same time we had a similar case in the north of Scotland, reported by a young couple who had been fitting spotlights in their newly refurbished bedroom. They had cut a hatch into the ceiling to feed electrical cable across the roof space but it kept getting caught on something. They prodded and pushed and eventually a bundle of clothing fell from the space in a great cloud of dust and debris. It was a dress from the 1950s, and wrapped inside were the desiccated remains of a newborn baby.

Almost predictably, the child was full term, as evidenced by the length of its clavicle, and there was no evidence of a cause of death. Sometimes it is not in the public interest to initiate a full-scale investigation. Who do you investigate? Who do you prosecute? Unless the property is still owned or occupied by the same family, how do you establish who lived there at the time? Finding anyone still alive who would have any useful information to offer, let alone tracking down any individual who might admit to being responsible, is almost impossible.

But there are times when we are able to reunite an infant who died at birth with their name, even many years later. It was the clavicle again that assisted us in the case of a baby whose body was found in very sad circumstances. A woman walked into a police station in the Midlands one day and informed the desk sergeant that, twenty years before, she had miscarried a child in late pregnancy. She was unmarried and, having concealed her pregnancy from everyone, she felt she could never confess to the stillbirth.

She told police that she had given birth to her daughter alone on her bathroom floor. She said that the baby had been born dead and never cried. She cut the umbilical cord and wrapped the baby in newspaper.

When the placenta was expelled, she had thrown this in the bin. But she did not know what to do with the baby. She could not bear to be parted from her daughter. However, burying her at home was not an option because she lived in rented accommodation, where she knew she was unlikely to be staying in the long term. She didn't want to find herself having to move on and leave her child behind. This reaction is quite common, and explains why the remains of miscarried or stillborn babies are often discovered secreted in unexpected parts of cemeteries or in suitcases, as well as in the dark recesses of houses. Since Violet's time, the population has become more mobile and it is unusual for anyone to spend their whole life in the same home.

This woman needed to find a way of giving her baby a decent burial that both concealed her and allowed her to be moved when necessary. She told the police that, after placing her little girl, still wrapped in newspaper, in an old pillowcase, she bought a very large metal plant pot to stand outside the back door of her house. She put compost in the bottom, laid the baby, in her makeshift shroud, on top, planted a bay tree in the pot and filled it with soil. She said that it didn't feel right to water the plant with her daughter there beneath it, so she let the tree die. But she kept the pot, and all of its contents, and it had moved house with her several times. At each new home, she stored it in a shed or cupboard to keep it "dry and warm."

At last, after carrying this secret for twenty years, and with it, no doubt, a large measure of guilt and anxiety, she felt she needed to unburden herself by telling someone the truth.

I was asked to attend the scene and assist with the recovery of the remains, if indeed there were any to be found. The plant pot stood some 60 cm off the floor and its circumference was of about the same size. As it was made of metal, we couldn't X-ray it, so we were going to have to perform a mini excavation.

The pot was taken to the mortuary and placed on a table where, layer by layer, the dry, dusty soil was removed using a paintbrush and a small garden trowel as a miniature dustpan and brush. We kept the soil to one side to be sifted later to make sure that we hadn't missed anything. The room was completely hushed, apart from the soft clicks of the camera photographing every step of the process. Everyone was holding their breath. A few centimetres down I could see a piece of cotton poking out and carefully removed the soil around it. This did indeed turn out to be a pillowcase, just as the woman had told police. It was intact and I was able to lift it out in one piece. I then cut carefully down the length of the pillowcase and unfolded the fabric to reveal the contents.

If there had been newspaper it was long gone now, but what was present was the perfect skeleton of a baby, its delicate, papery, desiccated tissues still visible where muscles would once have been, filling the spaces between the bones. The body remained largely articulated as the tendons and ligaments had dried and mummified, keeping everything in place.

The baby was still in the fetal position and the skull bones were misshapen, which is consistent with a vaginal birth. Every one of the tiny bones was recovered, photographed and analysed. The clavicle was 42 mm long, which confirmed that this was likely to have been a full-term birth. There was no evidence of a cause of death.

I don't believe Mum was charged with any crime, even though the offence of concealing a birth remains on the statute books. She was probably simply cautioned. The sadness of this case lies not only in the loss of a new life but in the trauma of a mother giving birth to a dead child in stark solitude and the psychological scars she had carried for two decades.

I gather that the little body was buried in a local cemetery. Although I don't know if Mum attended, I imagine, given the strong bond she had with her dead daughter, that she would have done, probably along with the family liaison police officers who would have been there for her throughout. Some people think of the police as hard and uncaring, but in my experience this is usually very far from the truth. In cases like this they share in the grief that is heavy in the air and show a deep humanity and sympathy for those they are trying to support.

As we were unable to establish a cause of death we could not say definitively that this was a stillbirth, but if it wasn't, this mother had surely been punished enough. Yet while the empathetic among us will view this as a tragic story of loneliness and bereavement, cynics will prefer to see an opportunist who murdered her unwanted baby and sought to conceal the death and her own guilt. It is unlikely we will ever know which interpretation is closest to the truth, but if she was guilty of any real crime, why would she ever have come forward? And I would rather live with hope than scorn.

The clavicle is unusual in that not only is it a reliable guide to the age of a fetus or a young baby, it continues to be useful in this regard right up to the end of the third decade of life.

As well as being the first bone to start to form in the fetus, it is one of the last in the body to finish growing. At the medial end (nearest the breastbone) is a plug of cartilage that gradually begins to ossify from around fourteen years of age (slightly earlier in girls than in boys). Eventually, as the cartilage continues to be replaced, the bone in the plug will fuse with the main shaft.

This starts to happen at around the age of sixteen, and a clavicle in a person anywhere between approximately sixteen and twenty-four can look as if it has a thin flake of bone glued to the medial end of the shaft, a bit like a scab on a wound. (Anatomically, "medial" denotes the area nearest to the middle of the body while "lateral" describes the region furthest away from the midline.) As completion of the fusion may not occur until the mid-twenties, it gives us a precisely defined range of possibilities: under fifteen, between fifteen and twenty-five or over twenty-five. So it is one of the first bones we will look to when we are trying to establish the age of either a child or a mature adult from their skeleton.

Although the clavicle is prone to fracture it is quite a resilient little bone, whether subjected to burial, exposure to the elements or fire. It owes its hardiness to its dense cortex and to the joint being tight with the sternum, which gives the medial end a certain amount of protection. It was this characteristic that provided an important clue to what may have happened to Marcella, a nineteen-year-old sex worker who went missing in the Midlands.

Marcella was the mother of a nine-month-old daughter. According to those who knew her, she continued in her high-risk line of business in order to support her little girl. One evening, Marcella left her daughter with a babysitter to go out to work and took a taxi into the red-light district of the city. She phoned the babysitter several times to check that all was well. The last call was made not long after 9 p.m. When she did not return to collect her child at around 11 p.m. as she had promised, the babysitter rang Marcella's mother, who contacted the police to report her missing.

Having checked all the local hospitals, to no avail, the police considered four possible explanations. First, that Marcella had chosen to abandon her daughter. Second, that she was being held somewhere against her will. Third, that she'd had some kind of accident and was lying somewhere injured, or worse, as yet undiscovered. Fourth, that she was dead as a result of foul play. As Marcella was a conscientious mother, they did not believe the first proposition was very likely. That meant the three other alternatives needed to be followed up as a matter of urgency.

Marcella's fellow sex workers were interviewed. Initially reluctant to give out names, descriptions or number plates for their regular clients or unknown kerb-crawlers, once they realized the seriousness of the situation, they helped police to very quickly narrow down a list of likely suspects. The police were left with two names to prioritize, one of which was of significant interest to them.

Paul Brumfitt had already served fourteen years in prison for two murders: he had first clubbed a shopkeeper to death with a hammer and then, while on the run in Denmark, he strangled a bus driver. He had also wounded a pregnant woman with a candlestick. He claimed that an argument with his girlfriend at the time had triggered this killing spree. With psychiatrists finding no evidence of mental illness, he had been freed from prison on licence and proceeded to rape a sex worker at knifepoint on two separate occasions. He was currently on bail for these offences. Might Marcella have become another of his victims?

As part of his rehabilitation, since his release from prison Brumfitt had been working as a gardener and park-keeper for the local council. The police, aware that he was renting a small lumber yard, focused their investigation on the yard and on the flat where he was living. At his home they found a small amount of blood which matched Marcella's, but it was insufficient to justify charging him with murder. At the lumber yard they discovered the remains of a very large, free-standing bonfire, on a piece of ground that had obviously been used to burn all sorts of different materials over a long period of time.

When clearly defined layers are seen in a fire, it indicates that there have been consecutive fires rather than one big blaze. If a fire is stoked frequently it produces a more homogenous ash. This one would have to be searched methodically, layer by layer, to try to establish what had been burned there and where within the fire it was located.

Finds from nearer the surface of the ash are, of course, likely to have been placed on a fire later than material retrieved from closer to its base. So it was really important that this fire was carefully deconstructed and its depositional history painstakingly recorded, a job that had to be undertaken by an expert forensic archaeologist.

Experienced forensic archaeologists are a rare breed in the UK but the police managed to secure the services of the best, Professor John Hunter. It was at this point that the police also contacted me, and I travelled down from Scotland to assist. All I had been told was that the police wanted me to look at some fragments of what they thought might be bone which John had recovered from the bonfire in the lumber yard. I also knew they were looking for a missing woman who they suspected had come to a violent end, and that their prime suspect was linked to the lumber yard in question, but I knew nothing more about Marcella herself.

John had systematically stripped away the burned debris from the bonfire, separately wrapping and labelling material from each layer and sending it to the mortuary, where I sifted through it. Each bag was opened and its contents spread out on the mortuary table to be examined item by item, much of it with a magnifying glass, to try to determine what might be present. Handling fire debris is dirty work. Everything is black or grey, so you need good eyesight and good light to make sure you don't miss anything. At the top of the fire was a lot of wood, some still unburned, that had clearly been used as its main fuel. Towards the upper section, some bone was found. It was not human, consisting for the most part of meat bones from food waste.

As I progressed down into the deeper layers, very small fragments of bone started to emerge. These were not obviously animal and were quite possibly human. They were grey in colour, which meant they had been burned for some considerable time, and so little of them was left that the likelihood of extracting DNA was close to zero. With several bonfires having taken place on the same spot, the ashes showed there had been repeated burning of the bones that reduced them into smaller and smaller pieces, perhaps pointing to an attempt to destroy them completely and ensure they could not be identified.

A set of house keys was also found in a lower layer of the fire. When tested by the police, they opened both the front and back doors of Marcella's home. But on their own, the keys amounted to only circumstantial evidence. Unless the remains could be attributed to Marcella, Brumfitt could not be charged with her murder.

Few of the bone fragments were bigger than a fingertip—and one of them actually was a fingertip. The bone was small, but I could tell that it was from an adult because the growth areas had fused. I identified another fragment as coming from the leg, from the lower end of the fibula, which forms the outside bump of the ankle. This told me that the growth plate, the area of growing tissue at the ends of the long bones, was fused, but that this had occurred only relatively recently. In females, fusion of this growth plate is normally complete by around sixteen to eighteen years of age. I also had a small section of alveolar bone from the jaw (the socket area, where the teeth sit), which was X-rayed to allow a forensic odontologist to compare our images with any radiography that might exist in Marcella's dental records.

And then there was the trusty clavicle. A piece of this, about the size of a thumbnail, had survived the sustained and intense heat of the fire, just enough to enable us to put the age of the victim at between sixteen and twenty-one. We could see that the sliver of bone at the medial end had started to fuse but that this was at a very early stage of development.

It turned out that what these tiny remnants of a human being were able to tell us fitted with the description the police had been given of Marcella. She was nineteen years old, in the middle of the age range established from the bone fragments, and she was a petite woman, under 5 ft (1.5 m). The police had been told that she often wore very high heels to make her appear taller. She looked so young for her age that it seems she capitalized on her appearance to cater for a particularly unsavoury type of client: one of her working outfits was designed to attract men who fantasized about sexual encounters with schoolgirls.

The police believed what happened may have gone something like this. Marcella was picked up by Brumfitt in the red-light district and he persuaded her to go back to his flat. There it is possible, given the suspect's previous offences, that he raped her at knifepoint. For whatever reason—perhaps she fought back, or to prevent her from identifying him—he stabbed her. He was already on bail for rape, so if caught, he would be sent straight back to prison, and he was no stranger to murder. This would explain the discovery of her blood in the flat.

Although we knew there had been dismemberment of the body, there was insufficient blood in the flat for this to have taken place there. Perhaps Brumfitt transported Marcella's body to the yard where, over a period of time, her remains were burned, piece by piece, along with her clothes and belongings. The animal bones that were found were perhaps merely the remains of meals that had been tossed on the fire, but they could have been added in a deliberate attempt to confuse anyone who might rake through the ashes.

Brumfitt was arrested. Initially he refused to answer any questions but finally he broke and admitted to the murder. However, he never elaborated on the details of the killing or the dismemberment. He subsequently received three life sentences for the two rapes and the aggravated murder of Marcella and was sentenced to life in prison, where he currently resides.

At the trial the judge accepted the identification of Marcella on the basis of three pieces of evidence: the testimony of the odontologist that the piece of alveolar bone matched a previous X-ray, the door keys found in the fire and our age confirmation, made primarily on what the thumbnail-sized fragment of charred clavicle had to tell us. Small wonder that this is a bone much loved by the forensic anthropologist.

◊

In contrast to the clavicle, the scapula, the second bone of the pectoral girdle, rarely provides much insightful information for forensic investigation. Although, again unlike the clavicle, it is quite difficult to break, it can be relatively easily dislocated at the shoulder because the upper limb is not firmly connected to the trunk.

This is a feature that has long been exploited in cases of torture. The method known as strappado, or reverse or Palestinian hanging, involves tying the victim's hands behind their back and suspending them from a rope attached to their wrists. With the shoulders in this position, the weight of the body frequently dislocates the humerus from the scapula. Because it is a lax joint, it can be rearticulated, which makes it ideal, from the torturer's point of view, for repeated assaults. Sometimes extra weights are placed on the shoulders to increase the agony. The pain is said to be excruciating and it can be fatal if the victim is left hanging for too long. The risk of death depends on the age and state of health of the victim, but strappado can lead to asphyxia, heart failure or thrombosis.

Quite apart from the danger to life and severe psychological effects, it can have long-term physical consequences such as loss of sensation (paraesthesia) in the skin of the upper limb, caused by damage to the nerves in the armpit region, or muscle paralysis, primarily as a result of injury to the axillary nerve.

The most important muscle to be affected is the deltoid, which covers the front, top and back of the shoulder area. As this muscle is the main one that controls our capacity to lift our arms to the side, a lasting legacy of strappado can be the inability to raise the outstretched arm to shoulder level. This is therefore frequently used by human-rights practitioners as a test to determine whether there is physical evidence to support the testimony of those claiming to have been victims of this form of torture.

It may be possible for the forensic anthropologist to detect the effects of strappado in a skeleton, provided the individual survived the original torture. Long-term damage to the nerves leads to muscle wastage, and it is likely that, in the regions where the muscles attach, in particular the deltoid muscle, areas of bone reabsorption will have occurred in anyone who was tortured in this way during their life. These marks, which are called enthesopathies, are effectively scars left by the attachments of tendons or ligaments that have been damaged through trauma.

While such inhumane acts might sound as if they belong in the past, unfortunately these techniques are still employed today to elicit information or confessions or to break the will of the victim or other prisoners forced to witness it. The human body is a marvel of engineering, but we know its limits. In the hands of those who choose to use such knowledge to push the body beyond them, it becomes a cheap and effective weapon.

The scapula owes its durability to its intrinsic sturdiness and to being protected by surrounding muscle. The origin of the name of the bone may be the Greek *skapto*, meaning to dig or delve, because of its spadelike appearance. Indeed, with minimal modifications, the shoulder blades of big animals such as cow, horse or deer were pressed into service by many ancient cultures as agricultural implements and used like hoes or trowels.

Although the scapula is not often pivotal to a forensic investigation, that doesn't mean we don't examine it in great detail. A stab or a bullet to the back may well leave its mark on the bone, and blunt-force trauma caused by a weapon like a baseball bat or a metal pole can fracture it. It is claimed that stress fractures can occur as a result of using axillary crutches. Sometimes diseases such as osteoarthritis or infections may be detected from the scapula and congenital or developmental anomalies, although rare, are occasionally reported.

The bone has its role to play in the confirmation of sex. In general, male scapulae are bigger than the female's and have larger sites for muscle attachment. Some say it can assist in establishing whether an individual was right- or left-handed. However, the main value of the bone is in age determination, especially in subjects between the ages of ten and twenty, when all the different parts that make up the adult bone start to come together to achieve its final formation.

In the fetus, the scapula starts to form up in the neck region before descending to its final resting position on the back of the chest wall. The congenital condition Sprengel's deformity is caused by the failure of the scapula to descend, resulting in one unusually elevated shoulder. Sometimes both shoulders can be affected. This is more common in females than in males: about 75 per cent of reported cases are females. It is linked to several other conditions such as congenital scoliosis. An omovertebral bone, a rare anomaly where the scapula is fused to the vertebral column by an extra bone, may be created by the ossification of the soft tissues that lie between them.

Right at the tip of the bony shoulder is a projection from the scapula, the acromion process, from the Greek for the top of a rocky outcrop (the same source as the word Acropolis). The tip of the acromion process starts to form in bone when we are around fourteen to sixteen, eventually fusing to the body of the scapula at about eighteen to twenty years of age. This is important for muscle attachment as it is a site of insertion for the powerful deltoid muscle. The deltoid forms the contour of the shoulder and controls its movement: when it contracts, the anterior fibres help it to flex the shoulder (bringing the arm forwards), the lateral fibres bring about abduction (elevating the arm to the side) and the posterior fibres support extension (pulling the arm backwards).

These are all actions involved in many sports, especially those that require power in the upper-limb muscles such as rowing, weight-lifting and gymnastics. If too great a strain is placed repeatedly on the acromion process by the deltoid in the young, the acromion may not fuse to the remainder of the scapula at the end of puberty and remain as a separate bone—the os acromiale. If this does happen it causes no pain or other problems in the majority of cases; in fact, its owner might never know they have it.

The human body responds in a variety of ways to the stresses placed on it, especially by repetitive activities, and bone can retain discernible echoes of an occupation undertaken hundreds of years in the past. When the wreck of the *Mary Rose*, King Henry VIII's magnificent flagship, was finally raised in 1982, the bones of about 180 of those who perished were recovered. The ship had gone down in the Solent on a warm summer's evening in 1545, in full sight of land, while leading an attack on an invading French fleet. All but about twenty-five of the crew of 415 were lost.

As expected, analysis of the bones showed them all to have been male, and mostly young—the majority were under thirty and some no older than twelve or thirteen. The ship also carried over three hundred longbows and several thousand arrows, so it is likely that there was a strong contingent of the much-feared English bowmen on board. Examination of the bones by an osteoarchaeologist, Ann Stirland, revealed a disproportionate incidence of os acromiale: it was present in some 12 per cent of scapulae.

The scapulae of modern-day archers who take up the activity as youngsters often show os acromiale on one side, particularly the left, as this is the arm most commonly used to brace and take the strain of the bow. It is therefore not unreasonable to surmise that many of the men on the *Mary Rose* would have been taught archery from a very early age and that the presence of the acromial ossicle (little bone) was a visible remnant of their strenuous training.

The human remains from the *Mary Rose* had all been taken to

Ann's cottage near Portsmouth for storage and analysis. This is not something that would happen today: the bones would be secured in a laboratory for safety's sake. But in those less regimented times I was able to sit with Ann in her dining room one glorious summer afternoon, with all the bones laid out on a table, marvelling at the remarkable preservation of these incredible pieces of history. The os acromiale was something she was very excited about and we spent hours trying to reunite these little lumps of bone with the scapulae from which they had originated by searching for the best fit, not always successfully. I felt really honoured to be allowed not only to look at, but to handle these remains, and the memory of that afternoon, passed in the companionable silence of scientific study, is one I cherish. Every time I see a documentary about the *Mary Rose*, it transports me back to that perfect summer and Ann's boundless enthusiasm, her copious cups of tea and a tremendous amount of laughter and wonder.

# 7

# The Pelvic Girdle

*"The pelvis is a literal gateway to evolution"*
Holly Dunsworth
Evolutionary Anthropologist

The second girdle in the body, the pelvic girdle, extends all the way around the torso from the sacrum at the back to the pubic bones in front. This is the junction where the weight of the upper part of our body is transferred from our spine to our hips, and from there down through our lower limbs to the ground.

Each of the innominates, the two paired hip bones of which it is comprised, is constructed from three parts: the ilium (at the back and at the top), the ischium (at the bottom) and the pubis (at the front). The ilium is the section that forms a joint with the sacrum at the back and it has broad, flat blades for muscle attachment. It is the ilium that has the prominent lump we can feel on our hips on each side. The ischium (specifically the ischial tuberosity) is the bit that we sit on. The pubic bones are at the front and articulate with each other in the midline behind the area where our pubic hair grows.

The ilium forms first, in the second month of fetal life, followed by the ischium at four months and lastly the pubis, at around five to six months. At birth, the pelvic girdle is comprised of twenty-one separate bones (fifteen in the sacrum and three in each of the two hip bones). The three bones on each side will eventually fuse together towards the end of puberty to make the single bone known as the innominate— rather incongruously, given that this literally translates as "no name."

Fusion takes place, at around five to eight years, between the ischium and pubis, so that by the age of eight both innominates are in two parts. Between eleven and fifteen years, the ilium and the combined ischium and pubis all come together in the cup-shaped acetabulum of the hip joint. Each innominate will finally be complete when we are around twenty to twenty-three, once the crest that runs along the top of the bone stops growing.

The innominate is a rich source of information for the forensic anthropologist. It may be of little value in helping with height or ancestral origin, but it comes into its own in establishing sex and age at death. It is said that, presented with a whole skeleton to examine, we will probably be able to assign sex correctly around 90 per cent of the time. But if you had to do this from just one bone, you would always choose the innominate, which can tell us what we need to know to get this right about 80 per cent of the time.

The innominate would also be the top choice for age determination, as it can assist us in coming to a sound decision right the way through from the early years into old age. Age-related changes in adults between twenty and forty may be seen in the surfaces of both the sacroiliac joints and the pubic symphysis between the two pubic bones at the front. Here the alterations can be both developmental and degenerative. These are well documented by research, enabling the bone to continue to give us guidance on the likely age of a deceased person well beyond their third decade.

The pelvis is divided by a well-defined rim into the false (or greater) pelvis above and the true (lesser) pelvis below. The false pelvis is so called because it is generally considered to be part of the abdominal cavity. This provides large, flat sites for muscle attachment and holds some of the abdominal viscera. The true pelvis beneath is a much tighter space, which houses structures such as the bladder, the rectum and the internal reproductive organs.

The rim separating the false from the true pelvis is known as the pelvic brim, or pelvic inlet. At the other end we have the pelvic outlet, bounded by the coccyx at the back and the ischial tuberosities to either side. The inlet and outlet are the gateways of the pelvic cavity, through which our soft tissues, such as our gut, nerves and blood vessels, pass. It is also the transit route for material that we wish to expel from the body: the products of our urinary tract, our digestive system and our internal reproductive organs (ejaculate in men, menstrual matter and, of course, babies in women).

That the female pelvis is so firmly associated with childbirth explains its particular value in the determination of sex from skeletal remains. It needs not only to retain its ability to perform its full-time functions—keeping our guts inside and allowing us to walk on two legs—but to be ready to accommodate the biggest thing ever to pass into the pelvic inlet and out through the pelvic outlet: a baby's head. Once that descends through the pelvic inlet, trust me, you sure as heck want to get it out of the pelvic outlet as swiftly as possible.

Until the hormones associated with puberty kick in with a vengeance—especially the primary female hormone, oestrogen—the pelvis is equally paedomorphic in both sexes, which means we are unable to determine the sex of a child from the pelvic bones. In general, whereas increasing levels of oestrogen change the shape of the female pelvis quite dramatically, the male pelvis retains its more child-like form and just gets bigger in response to larger muscle mass.

The reaction of the female pelvis to the effects of oestrogen is to prepare the girdle in a variety of ways for its role as a birth canal. For example, during pubertal growth, the back of the pelvis and the sacrum are raised, straightening the hook-shaped greater sciatic notch (through which the sciatic nerve passes from the pelvis to the lower limbs) to adopt a more obtuse, or open, angle. This creates a more spacious pelvic cavity with a wider pelvic inlet and outlet, aided by changes to the sacrum, which broadens in the female. Her pubic bones, which remain fairly triangular in the male, become longer and squarer. This helps to increase the size of both the pelvic inlet and the pelvic outlet. The ischial tuberosities will be further apart in the female than in the male. If you are in any doubt, look at the saddle on an old-fashioned bicycle. The manufacturers used to make them broader for women's bikes to suit the wider gap between ladies' ischial tuberosities.

These minor modifications to the female pelvis are all designed to work together to allow the successful passage of a fetal head. And most of the time, they do. However, when it comes to childbirth, considering that the pelvis is already jam-packed with all Mum's internal wiring, plumbing and viscera, there is still precious little room to squeeze a great big head through that tiny space. It is said that, on average, the female pelvic canal is an inch narrower than a baby's head, so something has to give if the head is going to get through the birth canal safely. The truth is that both mother and baby compromise just a little, because, after all, only an inch or so has to be found.

As the time for birth approaches, Mum's ovaries and placenta increase the production of a hormone called, rather appropriately, relaxin. This helps to rupture the membranes around the fetus and soften the cervix. There is some evidence that it also softens the ligaments that hold the normally tight pelvic ring together and so permits a little bit of movement. At the same time, since the bones of the baby's skull are not yet fused, as the head passes through the slightly loosened pelvic canal, they can ride over each other, squeezing the brain beneath. This is why it is not uncommon for babies to be born with slightly deformed skull shapes, which generally rectify themselves shortly after birth.

Quite often we find pits and grooves at the sites of joints in the pelvis, specifically in the joints between the sacrum and the ilium and between the two pubic bones at the front. In the past, scientists considered these to be indicators of childbirth: they even called them "scars of parturition." Some even went as far as to equate the number of pits present with the number of live births a woman had accomplished, one pit for each delivery. Time and research has shown this to be nonsense. If my Uncle Willie's brothers and sisters had all lived there would have been twenty-four of them. His poor mother was pregnant virtually her entire adult life. If a pit had been formed every time she had given birth, her pelvis would have looked like Swiss cheese.

While these pits are noted much more frequently in females, they do occasionally occur in males, too, so clearly they can't be fully explained by childbirth. When we do see them, they are nevertheless generally a good indicator of the female sex, although they are more likely to be scars caused by ligaments stretching at the joint surfaces than evidence that a woman has delivered a baby.

It is not unusual for fetal bones to be found within the pelvic cavity of a skeleton and it is something that a forensic anthropologist will routinely look for. Childbirth is a hazardous time for both Mum and baby and fetomaternal mortality is always a risk. There is also a rare phenomenon which merits just a little mention. A lithopaedion, Greek for "stone baby," can form either from a primary abdominal pregnancy or from a secondary abdominal implantation following an ectopic pregnancy.

The egg is usually fertilized high up in the fallopian tube, but if this occurs as it crosses the gap between the ovary and the tube, it can sometimes be deflected into the abdominal cavity. In an ectopic pregnancy, the fertilized egg does not get as far as the uterus and instead implants in the fallopian tube. Should the tube rupture, the embryo can migrate into the abdominal cavity. Alternatively, if the egg is fertilized before it enters the fallopian tube, it may fail to cross the gap between the surface of the ovary and the fimbriae of the tube and embed directly in the abdominal cavity.

The embryo is a genuine parasite, and provided it can implant successfully on to an abdominal surface, it can survive and develop outside the uterus, sometimes for as long as twelve to fourteen weeks. This is the stage when a fetus normally shifts its pole for placental implantation, and if a placenta cannot get hold of a sufficient blood supply, a function for which the uterus is specifically designed, the abdominal pregnancy will usually fail and the fetus will die. However, lithopaedia have survived beyond this age. The oldest we know of lived for thirty weeks.

Since the fetus cannot be expelled naturally—it has no way out—and may in some instances be too large to be absorbed by the mother's body, it begins to calcify. It is likely that the conversion into bone is an autoimmune response to protect the mother from infection in the event that the fetus starts to decompose. And so it is slowly turned into a stone baby.

The medical literature recounts fewer than 300 authenticated cases of lithopaedia and in most the mother was unaware that the stone baby even existed until it was discovered during a pelvic examination, often for something unconnected. Women have gone on to give birth to other live children without knowing they had a secret passenger on board. A stone baby can weigh as much as four pounds, yet in some instances, a lithopaedion has remained in the body undetected for forty years or more.

The pelvis is susceptible to fracture, especially from impact in a fall, crush injury or road accident. It is a common outcome when pedestrians are hit by vehicles. Collisions in which the knees slam into a car dashboard are a particular hazard: the femora can plough into the hip socket and break the pelvis into many pieces. This type of fracture can be very debilitating as there is a risk of nerve damage, which may result in incontinence and impotence. So please, don't sit in a car with your knees touching the dashboard. Move your seat further back and stretch out your legs.

Because the pelvis is a ring, a break in one part of the structure is frequently accompanied by a second fracture or further damage in another: these are known as unstable fractures, and such injuries, and their consequences, can be complex. When a person has survived them, the scars remain in the bones for the forensic anthropologist to find and as the fractures will almost certainly have required hospital treatment, there are usually X-rays, CT or MRI scans on record for comparison.

It is also not unusual for gunshot injuries to manifest in the pelvis. I was asked in both of the following separate cases to attempt to retrieve bullets from exhumed skeletal remains to try to shed some light on who may have been responsible for each shooting. These men had been dead for some forty years, but in both instances, the question of who had fired the gun was now of some importance to wider investigations. Both had been buried without postmortem examinations and without the removal of any ballistic evidence. This seems inexplicable today, but perhaps it was just a symptom of the time and place in which they lived.

The first victim was a young man of eighteen who had been standing on a street corner in Belfast talking with a friend when he was shot in the leg from a passing car—a classic drive-by shooting. He was rushed to hospital but died later on the operating table. The medical notes identified a ballistic entry wound but no exit, suggesting that the bullet may have stayed within his body.

As part of legacy investigations, a decision was made to exhume his remains and examine his skeleton for any ballistic evidence. He had been the first of his family to be placed in the grave, but three relatives had since been buried on top of him and the process of exhumation was going to be lengthy and complex.

The difficulty of this grim task was exacerbated by the weather. Exhumations always seem to be required when it is cold, dark and wet. It makes for a miserable scene as everyone huddles into tents to shelter from howling gales and lashing rain. You also know from bitter experience that the grave is going to become waterlogged and that you will soon be up to your knees in mud and water.

The most recent burial in the grave was that of a child, who had been interred in a cotton shroud that was visible very close to the surface. These remains were very carefully exhumed by hand using a trowel and placed gently into a body bag for reburial at a later date. A mechanical digger was then employed to strip away thin layers of soil until the top of the first adult coffin lid was uncovered.

At this stage, all that was needed was a quick jump into the grave to check the name on the plate, open the lid, transfer the skeletal remains into a body bag and hand out the disintegrating MDF pieces of the coffin. But by the time we got down to the lid of the second adult coffin a ladder was required. When a grave is deep it is difficult to manoeuvre in the tight space. For a lady of advancing years, and "wide in the beam," as my father used to say, it is always helpful to have a younger, fitter, thinner colleague working with you. Lucina, bless her, knows that she will always end up being sent down the hole.

The second set of skeletal remains were lifted from their coffin without incident and transferred to a body bag. The body bags, complete with their contents, were stored on site in what is known as a transportation coffin, which is really just an oversized wooden box, to await reinterment when the investigation was over.

The coffin of the young shooting victim was located exactly where the cemetery records said it would be, which, I can tell you from personal experience, is not always the case. We checked the name on the plate, removed the lid and transferred the remains to a body bag as the coffin was too rotten for us to try to lift it intact. A metal detector was passed over the coffin, and, once all the detritus had been removed, it was used again to scan the soil on the floor of the grave to make sure that no metallic pieces of evidence had been missed. Nothing was found.

The body bag was radiographed, using a mobile X-ray machine that had been brought to the cemetery. Everything was done in the presence of family members and their legal representative, to ensure that all aspects of the exhumation were open and transparent. For understandable reasons, there was considerable distrust between the victim's relatives and the police. The family had also engaged their own forensic anthropologist to monitor the operation. It was hoped that these measures would instil a spirit of co-operation and help with the healing process.

Some metal was detected. Each find was discussed with the relatives, their lawyer and their anthropologist, and each in turn discounted as pieces of funeral furniture or nails used in the construction of the coffin. Then one object gave off a metallic signature that was of interest, because it was associated with the bones themselves: specifically, with the pelvis. We would investigate this further at the mortuary the following morning. The body bag was then transferred to the security of the mortuary, accompanied by a police escort and family observers.

As the police drove us to our hotel in one of their vehicles, I admit to a complete brain-fade moment. Feeling warm, I asked why the windows wouldn't open. They looked at me as if I was joking.

When they realized I wasn't, they informed me patiently that you can't have bulletproof windows that open. The two features do not mix. It was a sobering reminder of the instability of Northern Ireland's recent history.

The next morning was bitterly cold, which made the chilly mortuary conditions particularly inhospitable. No amount of layering of socks or tops could get us warm. The body bag was photographed and opened on a side table and then, fingers freezing, we set about the first job in any skeletal evaluation: laying out the remains on a second table to make an inventory of what is present and what is absent. As each bone is placed in the correct anatomical position, a skeleton slowly takes shape and, from a jumbled bag of bones, the person they represent begins to materialize in front of your eyes. It always amazes the police and the legal observers that order can emerge from such apparent chaos.

As we reconstruct this human being, we are thinking about the features that indicate sex, age, height and ethnicity. We are looking for any anomaly, injury or evidence of disease in every single bone of the two hundred or so we will handle. All of these bones, and especially the innominates, confirmed this was the body of a young man who would have been in his late teens or very early twenties when he died. We identified unhealed fractures to the front of his ribs and sternum, which were in keeping with the hospital records stating that his chest had been opened and direct heart massage performed in an effort to keep him alive. There were also unhealed fractures to the bones of his right hand. The medical reports noted an entry wound in the region of the right groin, and accordingly the right innominate showed fracturing of both the superior and inferior parts of the ischio-pubic region—a double fracture that isolated the right pubic bone from the rest of the pelvis.

The metallic object we had seen on the X-ray the previous day was embedded within the inner surface of the pubic bone of the left innominate. The pattern of fracturing suggested that the ballistic projectile passed first through the man's right hand, breaking several bones there, then entered his right thigh and travelled upwards to fracture the right side of the pelvis before losing most of its momentum and finally lodging in the left pubic bone.

It was not our job to remove or analyse the bullet, just to find it, and that was what we had done. It was taken out by the pathologist, using plastic tweezers, sent off for analysis and that was the last we saw of it. I have heard no more about how that case has progressed, if indeed it has. Our brief was first to identify the coffin and lift the remains and secondly, to record, retrieve and present the evidence insofar as it related to establishing the trajectory of the bullet and its final resting place. Our task was complete.

The second case was very similar: a forty-one-year-old man, in the same part of the world, also shot in the right leg. He had been rushed to hospital but his leg could not be saved and had to be amputated. Two days later, he died of medical complications. Again, the medical records indicated that there was bullet entry but no evidence of exit, suggesting that a ballistic projectile might remain within the body.

An exhumation had already taken place—I was not present this time—but there had been a little complication with this investigation, in that the victim had been buried twice. His family had wanted to relocate him from the cemetery in which he had originally been interred to one closer to where they lived. With the body already having been moved once, the police had not been hopeful that the bullet would now be found, but it was. It had been picked up by a metal detector in the jumble of human remains and coffin detritus and removed by the pathologist for analysis.

My colleague Rene and I were summoned first to a lock-up in the Police Service of Northern Ireland headquarters, where the wood from the coffin and associated artefacts had been stored overnight, to look for any other evidence that might be of interest. The bones had been transferred to the mortuary, where we would examine them later in the day. On our knees on the concrete floor of the lock-up, we sifted through a small mountain of graveyard flotsam and jetsam, which included large planks of wood, religious iconography, metal nails, escutcheons, pieces of cord, scraps of cloth, dirt and stones.

The only thing we discovered was a finger bone, a metacarpal, that had been missed. This was bagged and tagged for us to bring with us to the mortuary. The probability was that we would find this particular bone missing from the skeleton. If we didn't, there were going to be some serious questions to be asked and answered.

At the mortuary we were relieved to find that the metacarpal was indeed missing from the skeleton and that the one we'd found was the right size to belong to the victim. It was clear, too, that the right leg had been amputated below the hip joint, which corresponded with the information in the medical records. This reassured the family's lawyer that we had the correct body.

Both pubic bones had been fractured and separated from the rest of the innominate bone, most likely as a result of the impact of the ballistic projectile. The bullet itself had, of course, already been removed, but the right pubic bone showed a starburst fracture that was consistent with the projectile, by that stage travelling at a lower velocity, becoming lodged in the top of the bone. The fact that there was no healing of the fractures supported the likelihood that they had occurred around the time of death. Again, that was our job finished, and we delivered our report to the police.

Is it a coincidence that these two cases from around the same time were alike in so many respects? Both victims were men, shot just once, both in the right leg, both had bullets embedded in their pelvic bones. Both died from their injuries and neither was given a postmortem examination, even though entry wounds, but no exit wounds, had been reported. Whether any or all of this was coincidental or evidence of a pattern of behaviour was for others to ask and answer.

All manner of items may be found in the pelvic region, so it is an important area of the body to examine carefully. And not only the remains but deposition sites, too, should be carefully searched with metal detectors. Genital piercings are commonplace, and a huge range of metal bits and pieces can be used to pierce or modify the genitalia of both males and females. Probably the most unusual I've come across was a scrotal ladder, which involved eight rings inserted into a row of piercings along the midline of the scrotum, with something resembling a very big safety pin running through them, all connected to another ring at the tip of the penis. I can only imagine the pain, but from a forensic anthropologist's point of view, it was certainly helpfully distinctive.

Other foreign bodies that regularly turn up include bladder stones, an assortment of intrauterine contraceptive devices and suspect packages associated with illegal activities such as drug trafficking. In one case we even retrieved a toothbrush from the anal canal. No matter how many questions we asked ourselves, we never came up with a plausible explanation for that one.

8

# The Long Bones

*"It is therefore indisputable that the limbs of architecture are
derived from the limbs of man"*

Michelangelo

Artist, 1475–1564

The long bones of the human upper limb and those of the lower limb are directly comparable, in other words, they are homologues. This is not surprising given that we were originally a quadrupedal animal. But millions of years ago, many species of tetrapods found they could forgo equivalence of power in all four limbs as long as they retained it in their hind limbs. This freed the forelimbs to do other things. Think of a squirrel grasping a nut or climbing a tree. In general, when there is modification of the forelimb in terrestrial animals, it tends to be shorter than the hind limb. Which is why you can't wear a cardigan on your nether regions without looking like a rapper with a hanging gusset. Every child has tried it.

It has recently been suggested, on the basis of something called a "constraint hypothesis," that kangaroos have small forelimbs because they are born at such an early stage of fetal development that the forelimbs need to be well developed to successfully make the perilous climb to their mother's pouch. This is critical to survival and the forelimb is therefore "constrained" by the early maturity needed to ensure that it meets its primary function. The hind limbs remain unconstrained and so can continue to grow.

This hypothesis has also been used to explain why there are no marine or airborne marsupials. A hotbed of scientific debate has long surrounded the reasons behind the "vestigial" upper limbs of the therapod Tyrannosaurus Rex. Maybe these were a grappling hook for mating or pinning down prey, or perhaps they were used as a lever to help it get up from a prone position. We may never know.

When describing our limbs after the point, about 4 million years ago, when our species decided to stand up and walk on two legs, we refer to them as upper and lower limbs rather than fore and hind limbs. Our upper limbs connect our body to our hands so that we can perform complex tasks and interact with the world, while our lower limbs connect our body to our feet so that we can move.

Anatomists are very specific about naming the parts of the body to ensure there is no ambiguity about which bit they are talking about. The part of the upper limb closest to the trunk is the arm and its equivalent in the lower limb is the thigh. Each contains a single long bone: the humerus and the femur respectively. The section of the limb furthest away from the trunk is the forearm in the upper limb and the leg in the lower limb. There are two bones in each of these segments. In the forearm these are the radius (on the thumb side) and ulna (on the little finger side), and in the leg they are the tibia (big toe side) and the fibula (little toe side). The radius in the forearm corresponds to the tibia in the leg and the ulna to the fibula. These six bones on each side—humerus, radius, ulna, femur, tibia and fibula—are known collectively as the long bones.

In our early years, our long bones grow at a fairly predictable rate and we can therefore say with reasonable accuracy what height we expect a child will be at two or ten years old. After that we start to lose our confidence. There will be a largely unpredictable growth spurt during puberty, an unpredictable event in itself in terms of when it will start and when it will stop. Once our long bones have finished growing (usually by around the age of fifteen or sixteen in girls and eighteen or nineteen in boys), we will have reached the maximum height we are ever going to be.

Our upper and lower limb bones increase in length pretty much in harmony with each other and on both sides, so that we don't end up with a really long right limb and a short left one, or with long arms and short legs or vice versa. Provided, that is, everything develops normally.

Those of us of a certain age will remember the effects of thalidomide, a drug manufactured in the late 1950s and early 1960s, initially by a German pharmaceutical company. It was designed to relieve anxiety, insomnia and morning sickness in pregnant mothers. The tests performed on animal models could not have predicted the devastating effects the drug would have on human fetal development. Mothers were therefore not discouraged from taking the drug during their first trimester until it emerged that there was a direct correlation between the drug and certain birth defects.

The gravity and nature of these defects varied according to how many days into the pregnancy the mother was when she started the medication. Begun on day 20, for example, thalidomide was producing central brain damage in the baby. In the case of the long bones, it disrupted the growth of upper limbs when taken around day 24, and the lower limbs up to day 28.

Deformities included phocomelia, which manifests as significantly foreshortened arms, forearms, thighs and legs, but with the development of the hands and feet often being less severely affected. In the UK, the drug was withdrawn in 1961, the year of my birth, by which time it is thought that at least two thousand babies had been born with defects of one kind or another associated with the drug, around half of whom lived for only a few months. Those with survivable deformities adapted. I remember being in awe of the dexterity of one girl in my class in school who could write with her feet. She helped me to learn at an early age that with adversity often comes ingenuity and determination. These children also needed great resilience, as people can be very cruel to those who look different.

Given the rate of growth in the long bones throughout childhood, and the close correlation between height and age in children, it is no surprise that we can use the length of long bones to determine the age of a child. In an adult, we can use the same measurements to calculate height but not to estimate age. This is exemplified by the fact that we can buy a pair of trousers for a child based on their age but for an adult we will need their inside leg measurement as well as their waist size.

The long bones keep growing in length and width until we hit the end of puberty but if something happens that slows development, this interruption can often be seen in their internal structure. We add bone longitudinally. Growing bones have little caps on their ends and when the caps seal, growth stops. Any event that hampers that process means that bone doesn't get laid down normally.

Instead it is added in lines or bands of increased density parallel to the growth cap. This "stutter," which is visible on an X-ray, tells us that something has temporarily affected the growth of the long bones, although it doesn't tell us what it was. It may be something as simple as a childhood infection like chickenpox or measles or even a period of malnutrition. These marks, known as Harris lines, can be seen most easily on the distal radius or distal tibia, but they may be found on many other bones within the skeleton where there is a high volume of cancellous bone. Once the incident is over, normal growth resumes and over time the body will reabsorb these white parallel lines as if they never existed.

I was in a mortuary one day, looking at a mix of bones that had been brought in for investigation. It was fairly obvious that they were all animal and, having confirmed that, I prepared to make a swift exit from the room, where another postmortem was underway.

The body being examined was that of a young boy of no more than ten or eleven years old, who, the pathologist confided, had almost certainly hanged himself. Suicide in children as young as this is, fortunately, extremely rare and his family and friends were apparently finding it incredibly difficult to come to terms with this explanation as there had been no sign of any illness or anxieties that might have been troubling him. He seemed normal, he seemed happy and he'd had his whole life in front of him. The police said he was from a "good" family and that there was no evidence of any form of abuse, psychological, physical or sexual.

The pathologist popped an X-ray of the boy's upper limb bones up on the screen and then an image of his lower limbs. He was looking for fractures, current or healed, to see whether there might be any history of physical abuse. I remember saying, uninvited, "That's interesting," as I noticed three or four very clear Harris lines at the lower ends of both the radius and the tibia. The spaces between these lines, which showed that normal growth had resumed for a while before being interrupted again, suggested that some kind of disturbance may have been repeated at intervals.

The pathologist asked what I thought these might mean. I could not help, except to say that perhaps something like recurring illness might be a possible answer. I never imagined for a moment how the case would unfold, and indeed I would never have known, had it not been for the pathologist recounting the story to me afterwards in a bar at a conference.

The police talked to the family and their GP and established that the child had suffered from no obvious or recorded repeated episodes of ill health or anxiety. He had taken his own life just before his mum and dad were due to go on holiday and questions were asked about whether this might be relevant. His parents explained that, because they ran a seaside hotel, they were often not able to get a break during the school holidays so, every year for the past five years, they'd got away for a few days on their own in term time while Grandad, Dad's father, came to stay to look after their son. That was when the child's father broke down and revealed that his own father had abused him as a child. He had believed all this was in the past, but he now feared that history had repeated itself and that perhaps Grandad had been abusing his grandson. The grandfather was interviewed by the police and, after indecent images of child sexual abuse were found at his home, he eventually admitted that this was indeed what had occurred.

The lines we could see on the X-rays may have been the boy's body responding every year to the fear and the stress of anticipating his grandfather's visit and what he would have to endure in his parents' absence. On the last occasion, perhaps he had been so distraught that he had taken his own life at the end of a piece of rope rather than face the trauma again or share his dark secret with anyone.

The little boy's dreadful story was discovered too late to help him and only unravelled at all thanks to the testimony of some little white lines on the X-rays of some long bones. If I had been involved in the case, could I have said in evidence that the stress of abuse had caused the Harris lines? No, I could not. But their presence had been enough to lead the police down a particular route of inquiry that had resulted in an explanation, a confession, a conviction and the destruction of a family. At times the truth is very painful and its impact devastatingly wide.

The passage of time and age give us the perspective to reflect more dispassionately on our own lives and on how, in our childhood, we might have laid down our own traumatic memories in our bones. Biological healing and remodelling may remove the physical evidence, but the mental scars are much harder to erase.

I have often wondered if my tibia or my radius developed a telltale Harris line or two when I was nine. If they did, I know now that they were probably eradicated by the growth and regeneration of my bones by the time I was a teenager, and all physical proof would have been gone. My mental Harris lines will remain with me for the rest of my life, but I have learned to live with them in peace and accept them as a part of who I am.

It was a sunny day, one of those carefree days during the school holidays when a child is blissfully unaware that something is about to happen that will change their life for ever. My childhood had been sheltered and happy, and I had no concept that there were people in the world with malicious intent in their hearts.

At that time my parents ran a hotel on the shores of Loch Carron, on the west coast of Scotland. I remember walking round the back of the hotel, past the door to the public bar. I was heading for the hotel kitchen and the milk churn that had been delivered by train the day before, which was kept in a big fridge by the back door. In those days, fizzy drinks were a rare luxury but fresh, cold milk, sometimes so cold that little ice crystals would form in it, was irresistible—the perfect drink on a hot, lazy summer's day. I would grab a glass from the shelf as I passed and fill it to the brim, using the metal ladle that hung from the lip of the churn.

There were always tradesmen in and out of the hotel and that day some boxes of fruit and veg were being delivered. I recognized the lorry driver because I had seen him many times before, although I had never really engaged with him. He had never seemed particularly friendly. My mind focused on the cold drink I was going to fetch, I thought nothing of brushing past him on the pathway—until he grabbed me by the arm and pinned me against the wall with such force that my head cracked against it and I could feel the pebbledash pressing into my shoulder blades.

He told me that if I made a noise I would be in so much trouble with my parents. If I close my eyes now, I can still feel his vice-like grip around my wrists. I remember the searing heat of the pain as tissues ripped and I remember a scream forming somewhere deep inside me that rose up through my body like a head of steam with no outlet for escape. To this day, I have a fearsome tolerance of pain and a tendency to endure it soundlessly.

When he had finished, he brought his face down to mine. I can still recall the stench of his breath. He said I was to blame, that I was dirty and disgraceful. He said I had to keep what had passed between us a secret because if I told anyone I would never be believed. It would hurt my mother, she would call me a liar and she would never forgive me.

I remember the sensation of warm blood trickling down my legs and a crushing feeling of shame mixed with fear as I ran up the back stairs to the bathroom on the first floor and locked the door. I stripped all my clothes away. I had to get clean so that nobody would ever know. I had to keep the secret. I tried so hard to wash the blood from my clothes so that my mother wouldn't see it but I couldn't get it all out and I began to panic. I realized I would have to "lose" them and think up a lie if my mother asked where they were. He was absolutely right: I was a liar.

I ran a warm bath and I remember the shock of pain when I lay down in the water. I hadn't been expecting the bubble bath to hurt. I lay there, alone. Traumatized but in control and thinking fast. I was not sure what had actually happened but, whatever it was, I was certain it was wrong and utterly convinced that I was guilty of something bad, which I could never share with anyone because I would be in awful trouble. I couldn't cry. I chose to own both the physical and mental pain. I grew up that day. I may have gained a Harris line or two but in the process I lost my childhood.

In many ways, it was a reflexive decision to leave those days of innocence behind me. With my friends I took on the role of the "sensible" one, the mother figure, the quiet one, the introvert and the thinker, and I carried the secret for nearly ten years, never uttering a word to a soul, trying hard to protect myself and those I loved from what I had done wrong. But then one day my mother threw one of her resentful comments at me—"Do what you want, you will anyway"—which was her way of rebuking me for my remoteness and self-sufficiency. By now I was a young woman. I decided it was time to tell her the truth.

Then came the second wave of pain and the chilling acceptance that he had been right all along: she didn't believe me. And she was clearly hurt. She accused me of making the whole thing up. Looking back, I think it was more a case of her not wanting to believe me. It was easier for her to tell herself that I was lying than have to face the ugly truth that I had been violently abused at such a young age and had chosen to lock it inside me for all those years rather than take her into my confidence. To be honest, I don't think she would have coped either way as she never acknowledged the pain of life.

What she said next was perhaps a clue that her reaction was born of defensiveness. While still refusing to accept that this had really happened to me, she made an oblique attempt to find out who was responsible. She threw a name at me and said that if this were true, it was probably him. That hit me very hard. The man she named had never been anything other than kind to me. A good man, a kind man, a flirt, a funny man who liked a good drink, but he had never hurt me. I was enraged on his behalf that he could be so glibly accused of someone else's heinous crime. It instilled in me an early awareness, even if I didn't at the time appreciate its full implications, of just how easy it can be to accuse someone wrongly, thereby setting up a chain of events that may ultimately destroy their life.

The response of the second person I told couldn't have been more different. I was a young woman and he was a much older man, a police officer. Jim tried to persuade me to identify my attacker so that he could be brought to justice. I simply couldn't. There was no evidence, it would be his word against mine, and I just could not bear to relive every sordid detail with strangers who might judge me.

But my policeman was what I needed: a father figure who was gentle, kind, caring, patient and understanding. Many did not understand our relationship, and most disapproved of the twenty-five-year age gap, but Jim was the one who held my hand and my heart until I healed as much as I ever would, and I will always be grateful for his genuine love and care. He died a couple of years ago at the good age of eighty-two. I wish I could have seen him just once more to tell him what a difference he made to my life.

I have found myself able to be less cautious about acknowledging my experience as I have grown older. The man responsible is probably long dead, my parents have passed away and can no longer be hurt, and I have accepted that the guilt was never mine to own.

I spoke of it in public for the first time while being interviewed by Ruth Davidson, the former leader of the Scottish Conservative party, a tremendously caring and compassionate woman. I was astounded to find that I could talk quite openly, calmly and rationally to her about something that had been locked inside that box in my head for so long. I wish I'd had more courage when I was nine.

My husband has been my best psychiatrist and counsellor over the many years we have been together, but the final healing is here, in these written words, almost exactly a half century later. Sharing my experience in this way is a conscious decision, and I do so with a salute to my long-lost, but never forgotten, Harris lines.

Ruth asked me if the work I do on paedophile identification stems from my past. I had to think long and hard about this but I am certain that it does not. I didn't become involved in that forensic area until I was well into my forties and already a longstanding wife and mother of three. The images I have to look at are, of course, distressing, but I do so with a detachment that confirms to me that it is work, not a personal crusade. In my job I have seen the results of all manner of human suffering and in order to do it effectively, you have to be able to compartmentalize, and to focus on the life stories that both dead and living bodies have to tell you, at all times keeping them separate from your own life. As a head of CID once told me: "Don't own the guilt. You didn't cause it and you are not responsible for it."

Where my personal experience does have a bearing is in reminding me of the damage that can be done by those who accuse wrongly, or without evidence, or out of malice, and who may in the process ruin the life and reputation of an innocent person. So perhaps my sense of justice has its roots back in that dark and lonely childhood place, inhabited now by nothing but my memory. But I genuinely believe that the whole ethos of forensic science is to be unbiased, and what we strive to achieve is to see the right people on the right side of our prison bars. You are innocent until proven guilty by a jury of your peers, and that is the way it should be.

Forensic anthropologists know that the long bones of the limbs can be important in the analysis of human remains, yet they are frequently overlooked by other professionals. When they do take centre stage, it is often because there is no other part of the deceased available for examination.

One such collection of limbs was discovered, ironically, by a police diving unit on a training exercise. Police officers working in all specialist areas, including mountain rescue and searching for and recovering bodies, train regularly to maintain and extend their skills, and this unit was diving off a pier on the shores of Loch Lomond.

On their first dive of the day, they retrieved several packages wrapped in black plastic bin bags which, not surprisingly, they assumed had been dropped into the loch by training staff for them to find. However, once on land, they quickly realized that these were nothing to do with any training exercise.

Inside the bin bags were real human body parts. The first to be uncovered was a severed hand, followed by another hand attached to a piece of forearm, then a foot and partial leg and, finally, a section of thigh. The officers immediately switched from training into operational mode.

Further dives turned up all of the limbs, but as yet no head or torso. These two areas of the body are critical as they frequently bear the evidence of manner and cause of death, as well as being more likely to aid identification. The divers would keep searching.

I was called to the mortuary to assist the pathologist in extracting what information we could from the dismembered sections of the upper and lower limbs. Identifying the victim was the priority, and any evidence we could find at this stage might put the police investigation on the front foot.

Fingerprints and DNA did not match any records on the police databases, which meant that this was unlikely to be anyone with whom they'd had previous dealings. But as the remains were relatively fresh, I was able to advise that if this was someone listed on their missing persons database, it was likely to be someone whose disappearance had been reported only recently. This enabled the police to swiftly narrow down their search for individuals whose descriptions might match our body. And it turned out that the limbs had indeed been in the water for only one or two days.

I was able to establish that the remains were those of a male, and that he had dark hair. This was evident from hair patterning on the forearms, hands, thighs, legs and feet. I could estimate his shoe size and calculate his height at just over 6 ft (1.8 m). His long bones had stopped growing but the fusion between the different parts was relatively recent, so he was likely to have been in his late teens or early twenties. Even though the hands and feet had been severed, which clearly pointed to murder and dismemberment, we detected extensive chafing marks. Had the young man been restrained and struggled forcefully? It was possible that the motive for dismemberment was in part to attempt to conceal these marks.

A hit came back from the missing persons database suggesting a possible name for the victim. Barry, who had been missing for just a few days, was eighteen, had dark hair and was 6 ft 2 ins (1.9 m) tall. DNA was successfully extracted from the limb muscles, and a comparison with samples from his parents confirmed their worst nightmare.

Barry's torso was found further down the loch some days later, but this offered no further clues as to either the cause or the manner of his death. Several days after that, a woman was walking her dog along an Ayrshire beach, many miles south of Loch Lomond, when the dog showed interest in a plastic bag lying below the high-tide line. A quick kick to check what might be in it revealed what looked to be a human head. A DNA comparison with the limbs confirmed it as Barry's. All the body parts had finally been recovered.

By this time, from what they had learned of the killer's modus operandi, the police had a strong suspect. It was accepted by criminologists and experienced police officers that William Beggs, a sexual predator who revelled in inflicting excruciating pain and torture, was well on his way to becoming a serial murderer. He had exhibited both the pattern of behaviour and the appetites associated with sadistic killers and being caught and imprisoned did not seem to deter him. It is likely that many of his early victims never came forward to report what they had witnessed or endured, through either fear or misplaced shame.

Beggs liked to pick up young men from bars and nightclubs and take them back to his flat. He may have drugged them. One of his victims spoke of waking up in agonizing pain to find Beggs cutting symbols into the skin of his leg with a sharp blade. Beggs told him not to worry, it would all be over soon. The victim was so certain that his attacker's intention was to kill him that he jumped, naked, from a second-floor window. If he was going to die one way or another, he reasoned that at least if he fell to his death they would find his body and Beggs would be caught. Against all odds, he survived and Beggs was duly arrested, convicted and sentenced to six years in prison.

Beggs had served his time but he had learned from his mistakes and would continue to do so, improving his technique and taking care to reduce the risks of being caught. In a classic example of the pattern seen in serial offenders, his behaviour escalated and his rituals evolved. For example, he took to handcuffing victims at both the wrists and the ankles, perhaps to enhance the sexual theatre but also to prevent them from escaping.

One night he picked up a young male student in a bar and took him back to his flat, where he manacled him before sexually assaulting him and, again, making cuts in his skin. He then cut his victim's throat. Beggs tried to dismember this body but evidently found it more difficult than he'd expected.

The human body is basically constructed of six parts: the head and torso form a midline axis, while the paired upper and lower limbs stick out from the side. The four limbs make a dead body a terribly unwieldy and heavy object to move, and difficult to hide. So when somebody decides to cut it up to make it easier to dispose of, separating it into five of its constituent parts is the most common approach. Dismembering a body into all six sections involves removing the head as well, which proves a step too far for some.

The inexperienced dismemberer, and let's face it most of us are, will probably attempt first to cut through the long bones. If they do, they will very swiftly find that this is an extremely difficult task. It requires the right tools, plenty of time, a suitable location and a good deal of stamina.

On this occasion, Beggs gave up and dumped the body to decompose in woodland, where it was found by a member of the public. He was arrested and found guilty of sexual assault, aggravated murder and a long list of other charges. Someone with this past pattern of behaviour should clearly have been identified as a risk, yet he was to serve only two years of his sentence before being set free because an appeal was upheld on legal technicalities.

It was after his release from this spell in prison that he encountered Barry, a popular teenager who was working in a local supermarket while he decided what he wanted to do with his life. He was considering a career in the Royal Navy. It was close to Christmas and he had been to his work's party. By all accounts, he'd had a good time and didn't want the evening to end, even though he had already had a lot to drink. A friend offered to give him a lift home but Barry decided to go on to a local nightclub. It was the last time he was seen alive.

Initially, his parents were not too concerned as they knew he'd been to the party, which he'd been looking forward to. They thought he had probably drunk too much and would be sleeping it off at a friend's place. But when he failed to come home the next day they started to become concerned and, unable to track him down through his friends, they finally reported him missing.

At the nightclub, Barry had somehow been befriended by Beggs and ended up back at his flat. There it is likely that he was drugged, handcuffed by his wrists and ankles, sexually assaulted and murdered. This time Beggs's dismemberment technique was more successful. He cut Barry's body into eight parts, removing his head. Removal of the head may have made a cut to the throat difficult to confirm as the evidence may have been obscured by the decapitation cuts. It is likely that he severed the hands and feet to obscure the marks of the mana-cles. Beggs wrapped the limbs and the torso in bin bags and dropped them into the loch. It was unlucky for him that this was precisely where the police divers would be training just a couple of days later.

He kept Barry's severed head for a little while longer. This he threw it into the sea from the back of a ferry to Belfast, which explained why it was found so far away. Shortly afterwards he fled to the Netherlands. He was extradited, brought back to face trial in the UK and jailed for a minimum of twenty years. With his sentence now approaching its end, there are, not surprisingly, concerns about the prospect of his release. Can someone demonstrating this pattern of perverted behaviour really be rehabilitated? I sincerely hope so.

Our ability in this case to determine sex, age, height, shoe size and hair colour from the limbs alone, along with our assessment of the length of time the remains had been in the water, gave the police sufficient information to target likely matches on the missing persons database and identify the victim quickly. That in turn led to a swift apprehension of the perpetrator. The features that can be established from the limbs alone might not amount to conclusive evidence of iden-tity in a court of law, but they can provide strong intelligence to focus the direction of an investigation. And they don't always have to be real limbs.

One dark November night, police were called to a flat in an inner-city block in response to a report of shouting, screaming and breaking of objects that indicated a heated confrontation was in progress. When they arrived, they found the flat in utter disarray and a man lying on the floor of the living room. The paramedics were unable to save him and he was pronounced dead at the scene.

There was a lot of blood on the carpet, furniture and walls and clear signs that the victim had been beaten severely around the head several times. A postmortem examination confirmed that the cause of death was multiple blunt-force trauma to the skull, resulting in extensive blood loss.

I was tasked with examining the skull, reconstructing the pieces and attempting to establish what kind of weapon might have been used in the attack. With the first PM already having taken place, the broken sections of skull had been extracted, and the skull cap removed with a Stryker saw, to allow for examination of the brain and its coverings.

When bone is wet, as it will be when an injury is perimortem, and the trauma has been violent, the pieces will not always fit back together perfectly, especially when the bone in question is from the skull and the layers of diploic bone have split. So it can sometimes take hours of fiddling to try to figure out where almost unidentifiable little fragments were once joined. We use a type of heavy-duty superglue to bond the pieces of wet bone together and if you are not careful you often find your gloves welded to the three-dimensional puzzle you are trying to reconstruct.

The pressure on you to come up with the first two pieces that dovetail is immense as everyone in the room looks on in the expectation that you are somehow going to accomplish this reconstruction in the blink of an eye. Isn't that what they do on the telly? Then slowly, usually once everybody else has lost interest and drifted away in search of tea and biscuits, you start to gain momentum as more and more pieces gradually come together. Only then can you start to interpret the trauma analysis that will allow you to determine how many blows occurred and in what order.

It was clear that this man had been struck around the head at least three times. The first blow was to the front of his head and the second and third to the left side, perhaps when he was already on the floor. They had been caused by a blunt instrument, probably made of metal as there were some sharp edges. There was a curved edge to one of the impact points that suggested something like a crowbar, but there also seemed to be a second, sharper, point of contact, more like the tip of a knife. So it was possible there had been two weapons, which didn't make a lot of sense to us at the time.

The explanation, which we heard about from the police after the murderer was found guilty and had started his prison sentence, was so bizarre that I think we could be forgiven for not managing to pin down the murder weapon at the time of the postmortem examination.

The dead man, Michael, had been known to the police. As he had worked as a gay male prostitute, they had feared that tracking down who might have been at his flat that evening would be challenging. But Michael's regular haunts were duly visited and his fellow sex workers questioned, and two of them mentioned a man they had not seen before. They said that Michael had gone off in his company at some stage that night, though neither could be sure exactly when. Their description of this man was fairly generic—until one of them happened to comment, almost as an afterthought, that they had jokingly dubbed him "the Captain" because he had a prosthetic right arm with a hook on the end. Apparently, his pub party trick was to take off his prosthesis and hang it from its hook over the side of the bar.

The police couldn't believe their luck. Not only did they have a possible suspect who was known to use the services of male sex workers and to have a violent temper when he had been drinking, but he wasn't going to be too difficult to find. "The Captain" was quickly located, picked up and taken in for questioning. He attributed his rages to the onset of PTSD after he was injured by a roadside bomb during his military service, which was what had led to the amputation of his hand and part of his forearm. His distinctive hooked prosthesis was removed for forensic investigation and blood matching Michael's was found around the cup area of the limb, where the hook slotted into the wrist-based mechanism.

As well as the stainless steel hook at the end of the prosthesis, there was a finger-like tine that lay along its inner surface, apparently designed to increase its gripping capability. Forensic experts were able to match the hook and the barb-like shape of the tine to Michael's skull injuries and confirm that the prosthesis was the most likely weapon to have caused them. In the wrong hands, artificial limbs can kill.

Sometimes it is not so much the bones themselves that help us with identification as the joints between them. Like the bones, the joints in our limbs mirror each other: the shoulder is homologous to the hip (both articulating with a girdle), the elbow to the knee (at the intersegmental junction of the limbs, where the range of movement is limited) and the wrist to the ankle (where the long bones articulate with the terminal appendage of either the hand or the foot respectively).

These are all freely movable (synovial) joints but there are differences. While the hip and the shoulder can move in all directions (flex, extend, adduct, abduct and rotate medially and laterally), the shoulder has developed an additional ability to exaggerate movement into what we call circumduction. This means that we can swing our upper limbs around like a windmill, which is something few of us could achieve with our lower limbs, regardless of our flexibility. I dare you to try.

The downside of this increased mobility in the shoulder joint is a much higher risk of dislocation in the shoulder than in the hip. To have a hope of keeping us upright, the hip joint must be kept tight to bear our weight and must remain stable, whether we are standing or in motion.

The knee and the elbow are very restricted in their range of movement, especially the elbow, where movement occurs only in a single plane of flexion and extension, which is why these are known as hinge joints. The knee has a little more flexibility, just enough to enable the femur and tibia to rotate slightly on each other to lock the knee when we are standing. This is a mechanism that assists in maintaining stable stance and the reason why you have that momentary collapse of balance when someone unexpectedly pokes you in the back of the knee.

The wrists and the ankles are largely comparable, with a range of movement that allows our hands and feet to adopt a variety of positions, in keeping with their respective primary roles of manipulation and locomotion.

Our joints take a fearsome amount of strain from repeated use over our lifetime, and nowadays can be routinely replaced or resurfaced. Over a quarter of a million joint replacements are undertaken on the NHS every year. Hip and knee replacements remain the most common, but shoulder, ankle and elbow replacements are on the increase. These surgical procedures leave their mark on the skin surface in the form of very typical scar patterns in specific locations. And in parts of the world where such procedures are subject to strict legislation, there is usually a requirement for any implant to carry a unique reference number, which should be recorded in the patient's medical records. So when we come across these, they should be fairly easy to trace. But life is never that simple.

There is no global register of such information and in countries where medical tourism is in the ascendancy—the top three for joint replacements are India, Brazil and Malaysia, where the cost is low and availability swift—records can be far from complete. A hip replacement carried out privately overseas may never appear in a patient's NHS records, and tracking down the hospital where it took place will be extremely difficult. In less scrupulous corners of this blossoming worldwide industry, where the supplier may not even provide their implants with unique number identifiers, many of them have been known to carry the same number, which makes identification pretty much impossible.

The ability to identify artificial hardware, and related scar damage to its host, is a necessary part of the modern forensic anthropologist's skill set. The dissecting rooms of most anatomy departments have a box of various orthopaedic implants extracted from bodies donated for dissection which are retained to teach students to recognize these objects if they come upon them, perhaps among decomposed remains. Other ironmongery associated with broken bones and surgical interventions includes plates, screws, wires, pins, rods, nails and washers. Sometimes cataloguing what we find can seem more like a DIY-store stock-take than an assessment of orthopaedic surgical attention.

It is critical that artificial bits and pieces introduced into the body are understood just as well as its natural components. This foreign material tells us what kind of medical treatment the person may have received, and that may be key to establishing who they were or what may have happened to them.

It is not surprising that long bones are prone to fracture as they are the means whereby we interact with our environment. So it is important to check medical histories where possible. Previous healed fractures or the presence of orthopaedic implants are a good indication that medical records are worth trawling through. It has just crossed my mind as I write that my father's hip replacement has probably ended up in a box at the funeral director's office, perhaps in the same one as my mother's replacements for her big toe joints. How odd to think that parts inserted into both of them may have ended up in the same jumbled box of spares. They were both cremated and, as these implants do not burn, they must at some point have been extracted by the undertaker. I never thought to ask.

As we have seen, fractures can play an important role in the intelligence and evidence gathered from a postmortem examination. So a forensic anthropologist must be able to determine whether they occurred before, during or after death. Premortem fractures do not usually have any connection with the death itself, except, in some instances, as evidence of the possibility of past abuse. They can often be linked back to medical records, and the time in the person's life when they happened can be estimated from the degree of healing and callous formation that has taken place. Postmortem fractures may also be inconsequential to the death, although they can help us to build up a picture of the methods used to dispose of, or conceal, a body. So it is perimortem fractures, those which occur around the time of death, and which may be related to the manner of death, that tend to be of greatest intrinsic forensic value.

What are the sorts of injuries that result in our long bones breaking? The humerus is not a bone that fractures very often, but when it does, it is usually due to a fall or a sports injury. Because the radius and the ulna are connected by a membrane (interosseous membrane), if one bone is fractured, the other may follow suit. Fracturing of the radius is frequently the upshot of a fall on to the outstretched hand, when the impact of the ground on the heel of the hand will be transferred up the radius and across the membrane to the ulna. This is known as a Colles' fracture, named after Abraham Colles, a noted early nineteenth-century Irish professor of anatomy, who wrote a treatise on the subject, and it is particularly common in falls among the elderly, whose bones are more fragile and may be weakened by osteoporosis.

The radius and the ulna are also sites we examine for defence fractures. If someone is raining down blows on your head, you are likely to raise one or both of your forearms to try to deflect the assault, which often leads to fracturing of the shaft of either the radius or the ulna or both. This was one of the injuries we saw in Harry, the little boy who died at the hands of his father, in Chapter 4. In these circumstances the break will be in a different place from one caused by a fall, and so it can be pivotal in distinguishing between an accidental fall and a criminal attack.

Postmortem fracturing of the radius and the ulna is commonly seen in those who have died in a fire. As the muscles respond to the intense heat, they contract quite extensively and the body assumes a pugilistic pose, with the limbs flexed and the fists clenched. The strain on the sites of muscle attachment can eventually fracture the bones at the wrist, especially if they are further weakened by fire damage. Identification of fragile, burned bone fragments is a very specialized skill and these will need to be retrieved before the scene can be cleared of debris.

The task of recovering the remains of one elderly gentleman who died in a house fire we attended was typical of what we encounter. He had lived alone and was known to like a smoke and a drink, and the police and fire service were reasonably confident that there were no suspicious circumstances.

In buildings where a fire has taken place, the electricity will usually have been turned off for safety reasons and so you will probably be working with battery-operated lighting, unless a generator has been brought in. You will also be wearing a face mask and goggles and navigating a monochrome moonscape of various shades of black and grey. The floor is likely to be covered in debris, especially if the ceiling, and sometimes the contents of the floor above, have fallen through. None of which makes it easy to locate tiny fragments of charred bone. The man had died in his armchair in his sitting room, surrounded by piles of newspapers and several whisky bottles, many of them containing urine. The ceiling had collapsed on top of him. The fire officers believed that the seat of the fire was in the vicinity of his chair, and so it is likely that a cigarette was the source.

As human bodies are made up of so much water, they do not burn terribly well and what we mostly see is scorching of the skin, especially where clothing has caught fire. Areas of the body that are uncovered tend to fare the worst, usually the head, hands and sometimes the feet. In this case, the man's feet were less damaged than they might have been because his slippers had melted around them, protecting them to some extent.

His wrists and forearms were more badly affected. It was the height of summer, and he had probably been wearing a T-shirt which left them exposed to the fire. We could see that the fire damage and trauma caused by muscle contraction had resulted in the fracturing of his radius and ulna, and although parts of his hands were still attached to the long bones, some of his finger bones were missing. Exactly which ones had to be established at the scene because it was our responsibility to find and collect them.

It takes many years in this job and a thorough knowledge of anatomy to be able to recognize burned fragments of bone, some no bigger than the nail on your little finger, and to assign them accurately to a particular part of the body. But it has to be done, because you absolutely cannot have body parts left behind to be stumbled upon later or, even worse, thrown out with the burned debris from the house when it comes to be cleared.

First of all the body had to be lifted from the melted armchair and removed from the room. This is not easy when the fabric of the chair has melded with parts of the body and you have to cut away remnants of fabric or foam padding to free it. Fire victims are often rigid and fixed in their boxer's pose, which makes them awkward to lift and to secure in a body bag designed to accommodate a supine figure with limbs extended. But once the victim has been extracted and laid out on a body bag in the room, it is easier to examine the ends of the limbs and to determine where the postmortem fracturing has occurred and which anatomical parts remain unaccounted for.

Having compiled your inventory, you keep this in your head while you search the debris for what is missing. In this painstaking way, we managed to reunite this gentleman with all of his fingers and the lower parts of the long bones of his upper limb, which were transferred to the mortuary with him in the body bag.

Femora, as has been discussed, are often fractured in car accidents, when knees are rammed into the dashboard of the vehicle. The bone is also commonly fractured in the elderly as a result of bone loss through advancing age, which makes hip fractures another hazard for older people. These can be caused by something as innocent as turning over in bed. Somewhere between 70,000 and 75,000 hips are broken every year in the UK, 75 per cent of them in women with an average age of almost eighty years. The link between hip fractures and mortality is strong. Did this person fall because their hip broke, or did they break their hip when they fell? In these cases, trying to say for certain whether the injury was premortem or perimortem can be very difficult.

My father broke his hip after he was admitted to a psychiatric hospital because of his advancing Alzheimer's. He was unstable on his feet at the time and the nurses were comfortingly honest when they told us that they could not be sure whether he had tripped or whether one of the other patients had knocked into him or pushed him. If you cannot tell the difference between the two in the living, it is no surprise that it can be almost impossible to be certain in the deceased.

The kneecap, or patella, is the biggest sesamoid bone in the body. The term "sesamoid," from the Latin for "sesame seed," is usually reserved for small, nodular bones that develop in the tendon of a muscle. In the case of the patella, which obviously looks nothing like a sesame seed, it is a bit of a misnomer.

There is another sesamoid bone that might be found around the knee joint: the fabella ("little bean"), which can form in the tendon of the lateral head of the gastrocnemius muscle at the top of the calf. It is estimated to be present in fewer than 40 per cent of people and is more common in elderly men than in any other group. So when we come upon a fabella, it is worth finding out whether antemortem X-rays are available for comparison, as its presence may have some relevance in helping to identify the deceased. The function of the fabella is not well understood. There has been a suggestion that it is an evolutionary feature which has re-emerged, perhaps due to a combination of genetic and environmental factors. That theory seems more than a little far-fetched to me, but then, I am not a geneticist.

The patella, which is named for the Latin word for a small, shallow dish or pan, is located in the tendon of the quadriceps femoris muscle on the front of the thigh, where it starts to form in the third year of life. Its purpose is to increase biomechanical efficiency of the knee. Its prominent position puts it at risk of fracture, usually from direct impact to the knee or as a result of falling on to the bone from a height. If the kneecap is shattered into too many pieces it can be removed, but these days surgeons are more likely to opt for reconstructive operations, generally involving wiring and tension-banding parts of the bone together. So forensic anthropologists are always on the look-out for bits of wire that might indicate orthopaedic intervention here.

As kneecaps are sensitive places with lots of nerve endings, they are a popular site for inflicting pain. Kneecapping is a deliberate injury meted out as a form of torture or punishment, usually by a bullet from a handgun and sometimes with an over-enthusiastic swing of a baseball bat or similar weapon. In fact, in many cases the impact misses

the kneecap entirely, either by accident or design, and instead fractures the lower end of the femur or the upper end of the tibia or fibula.

Kneecapping was used in Italy in the 1970s and 1980s by the
Brigate Rosse guerrilla group, and during the Troubles in Northern
Ireland, by both Loyalist and Republican paramilitaries, as a punish-
ment for a variety of transgressions. In the course of the conflict in
Northern Ireland, around 2,500 cases are said to have been recorded.
The ultimate punishment was the "six-pack": a gunshot to each of the
elbows, knees and ankles. More recently its use has been reported
by the Islamic fundamentalist organization Hamas and by police in
Bangladesh.

Accidental fractures of the tibia and the fibula are mostly seen in
sports-related contact injuries or in collisions with moving vehicles,
sometimes referred to as "bumper" fractures.

Fractures to the long bones of the limbs, and of the lower limbs in
particular, need to be set in such a way that the length of both limbs
remains more or less equal. If a patient is left with one shorter than
the other, it sets up all manner of compensatory anatomical changes
throughout the rest of the body. If your legs are of different lengths
there will be effects on the pelvis and the vertebral column as well as
on the limb itself. When we find evidence of a poorly set lower-limb
fracture, we can say with some confidence that the person is likely
to have walked with a limp, or at least an uneven gait, during their
lifetime.

We will all have variations in the symmetry of our long bones and
this can offer evidence of our "laterality"—the preference we show for
one side of our body over the other. This is generally judged in everyday life by whether we naturally opt to write with our right hand or
our left, but of course it relates also to which foot we choose to kick a
ball, and to a predominance of one side over the other in general. And
being a left-handed writer does not necessarily mean you will favour
the left hand in every activity. For example, the right hand may be preferred when eating or playing a musical instrument or a sport. Plenty
of cricketers bat or bowl with their non-writing hand.

However, about 90 per cent of humans are right-handed, and if your dominant side is the right, it is highly likely that your dominant foot will also be your right one. The degree of laterality varies between individuals, but it is highly unusual for someone to be truly ambidextrous, that is, to have exactly the same level of capability in both hands. Right-handedness seems to be supported by the intensity of motor and sensory cortex on the left side of the brain. Those who are right-limb dominant are therefore left-brain dominant for motor and sensory functions as the nerves are contralateral, crossing between their origin and their destination. Lateralization is not restricted to humans: it is widespread throughout the animal kingdom and has been identified in most primates, dogs, birds and rodents.

While right-handedness might have some genetic origin, it is likely that forcing small children to use the right hand, which was common practice in the past, and the fact that almost all the implements we use in everyday life are designed for right-handers, results in pressure to conform to right-side dominance.

Historically, the right side of the body was considered to be the "correct" one, and those who chose to write with their left hand were viewed as untrustworthy or even downright evil. This belief had its roots in the founding myths of many ancient cultures and religions and is reflected in the etymology of the words "right" and "left" in most languages. Right equates with proper, correct or straight; left with sinister, clumsy or weak.

There has been a considerable amount of scientific research on right- and left-handedness, including its genetic influences, indicators in the fetus and effects on birth weight, intelligence, income and many other matters besides. Results suggest that even at birth, there is evidence of dominance in the dimensions of the limb bones, with the right tending to be longer and more robust than the left. As an individual develops, dependence on one side causes a corresponding increase in muscle mass, and the difference in size between corresponding bones, right and left, increases.

The right humerus, for example, is frequently longer, wider and stronger than the left due to enhanced blood supply and muscle development. We can see this in our fingers: a ring will often be tighter on a right-hand finger than it is on the same finger of the left hand. In those who become left-side dominant, the bones on the right side of the body still tend to remain slightly longer, wider and more robust, but the difference between the two sides is reduced.

While we usually describe dominance in terms of our hands and feet, the real muscles of power that operate the upper and lower limbs are in the arm and thigh and the forearm and leg. So the differences in bone dimensions are more likely to be influenced by muscle mass in the long bones of the limbs than in the shorter bones of the hand or foot, where the tendons attach.

Although we can measure variations in the size of the long bones, it would be dangerous to claim that we can determine right- or left-handedness from bones alone, though that has not stopped some less than scrupulous anthropologists from doing so in the past. In the postmortem rooms of bygone days, we might have looked for callouses on the fingers as an indication of which might be the dominant writing hand but in today's world of keyboards and keypads, those signs are rapidly becoming irrelevant or obsolete.

Of course, the odds are in our favour if we suggest that an individual is likely to have been right-handed. In practice it is recognizing the more unusual left-handers that is more valuable for identification purposes. But offering such opinions is risky.

When dealing with decomposing or disrupted remains, tracing the story of bones, of what has happened to them and when, can, as we have seen, be a challenge. Long bones are a terrific source of calcium for scavengers and the marrow in the cavities is appetizing and nutritious. As a result, they are very likely to be gnawed at when exposed to wildlife. We can sometimes suggest the type of animal that has been preying on them, whether it be small rodent or larger carnivore, by the marks left by their teeth on the surface of the bones.

But not all bodies will be predated, and sometimes the absence of carnivore activity can be as confusing for investigators as its presence, especially if they are not well versed in animal behaviour. To understand animal activity, you need to look at their life cycle holistically.

Foxes, for example, are not indiscriminate feeders. At certain times of year, they may totally ignore a large carcass such as a human body in favour of smaller, bite-sized meals that are more abundant, easier to manage and perhaps fresher, if they can make a kill.

They can be fussy eaters and human remains in the mid-liquefaction stage of decomposition are generally unpalatable to them. We might think that the more putrefied the remains become, the more attractive they will be to all predators, but this is simply not true. Foxes will happily feast on a relatively fresh body and crack into the bones to reach the calorie-rich marrow. And once decomposition has passed the mid-liquefaction point, they will return to gnaw at the bones for the calcium benefits. But in between, they will often leave it alone unless food is in short supply.

One body, found on the edge of some farmland in the central belt of Scotland, showed surprisingly little indication of animal predation, given the proliferation of foxes in the area and its advanced stage of decomposition. The police speculated that perhaps the body had been stored elsewhere prior to being dumped at the edge of the field. But there was no evidence of foul play.

We knew that the attention of foxes would be influenced by when the body was deposited. Was it at a time of year when food was plentiful, or when there were no cubs to be fed? So when asked by the police to give our opinion, we were able to persuade them, backed up by the views of vulpine experts and gamekeepers, that, like other carcasses, human remains left out in the open do not always bear the signs of carnivore scavenging activities, and that the absence of these marks in this case did not necessarily point to the remains having been recently placed at the site.

The hands were missing, but this could be explained by the foxes' food-storing habits. As they don't know when they will next be able to eat, they conserve their food and will remove portable body parts such as hands first, for later consumption, and bury them elsewhere. This activity is called caching.

Missing body parts can often be tracked down by following the foxes' trails and looking for disturbed ground. Foxes can be very protective of their larders and tend to "scatter cache" as a bit of an insurance policy rather than bury all their food in one place, where it will be at risk of discovery by other predators. But other animals do, of course, find and pilfer from the cache, with badgers being the main culprits.

The body was eventually identified as that of an elderly itinerant man. It was quite possible that he had crawled under the bushes to sleep for the night and simply died there. And the missing hands? They were subsequently recovered some distance away from the body, each buried in its own little cache. The bones showed the telltale puncture wounds of vulpine canine teeth.

The skin and soft tissue covering the limbs, as well as the bones beneath, can aid with identification. The most common part of the body chosen for tattoos is the forearms in men and the shoulder or hip in women. As for the designs themselves, we like to think our inkings are unique but in reality most people go into a tattoo parlour and either pick one from a catalogue or ask the artist to copy something they've seen on someone else.

One young man, who wished to prove that he had what it took to become a member of a paramilitary organization, decided to make a video of himself dismantling and rebuilding a handgun to demonstrate his prowess with firearms. Adept as he may have been at stripping down and reconstructing a gun, he was evidently not the sharpest tool in the box in other respects, as he filmed the video in his own kitchen, which was a bit of a giveaway when the police became involved.

It seemed that he had at least watched enough forensic pro-grammes on the TV to know that he shouldn't let his face, or any distinctive clothing, be seen, or leave his fingerprints on any part of the firearm. So he wore nothing above the waist, shot the video from an angle where he was visible only from the neck down and donned a fetching pair of yellow washing-up gloves. But with the area between the top of his gloves and his shoulders bare, his film featured enough glimpses of the tattoos on his forearms for them to be identifiable.

I compared the tattoos in the video with those sported by the accused. The ever-popular "Madonna with a rose" on his left forearm could be seen above the top of his Marigolds. On his right forearm, his Celtic cross tattoo had been inked right next to a very prominent nevus (birthmark), and both the top of the cross and the birthmark were visible above the right glove. So we could match not only the tattoos to our bold hero but his birthmark, too, which led to him being detained at Her Majesty's pleasure.

Many of us have a susceptibility to the formation of moles, freckles or liver spots, collectively known as punctate pigmentation. These are manifestations of areas of increased melanin, the pigment that gives our eyes, skin and hair their colour, found in the stratum basale layer of our skin. As melanin absorbs ultraviolet light, these areas darken in the sun. They also increase with age. Because these marks appear at random, they are specific to each individual, which makes them useful when comparing the anatomy of a suspect or victim with images captured on film of a person alleged to be the same suspect or victim.

Most of the cases we take on that require matching of anatomical features in photographs or videos relate to sexual abuse, frequently involving children. Two young girls accused a primary school caretaker, Peter Ryal, of sending them indecent messages by phone and touching them inappropriately. The police arrested him and obtained his phones and computer. Among the many images he possessed they found a short video taken on his mobile phone that showed another female asleep in bed. Her bra was raised and her exposed breast had been filmed.

The police investigated further and interviewed a teenage girl who was friendly with Ryal and his wife Gayle. She confirmed that she had stayed at the couple's home, sleeping in the spare room, after having too much to drink one evening. She had been unaware of being filmed but said that the female in the video was her. Her face was not visible but she recognized her bra.

Gayle Ryal told the police that she was shocked by the images. However, when the case went to court, she changed her evidence and testified that she was the woman in the video filmed by her husband, claiming that they had been role-playing.

The court found itself at an impasse and the judge ordered a retrial, instructing the police, in the meantime, to seek an expert who could tell the difference between the anatomy of the teenage girl and that of Gayle Ryal.

They asked me to examine the video. It was of reasonable quality and, by breaking it down into individual stills, my team were able to map the pattern of moles across the shoulders and arms of the female subject. Then we did the same with photographs taken of both Gayle Ryal and the teenager and compared them with the map we had created from the material retrieved from the mobile phone.

The specific question put to us in this instance was clear-cut. We were not being asked to assess how likely it was that there was a match with either of the alleged subjects but, more straightforwardly, was the female in the video Gayle Ryal or was it the young girl? Gayle had a lot of freckles but no moles, and we could not match her to the photographic evidence. The teenager had moles (and no freckles), and in a formation perfectly replicated by those in the images. The answer to the question posed by the court was plain.

I was called to testify at the retrial. The judge became increasingly frustrated with Peter Ryal's lawyer, whose only line of defence, it seemed, was to attempt to prove that I was not qualified to establish the difference between a freckle and a mole because I was not a dermatologist. Eventually the judge shut him down. The point was, of course, that for forensic purposes, it didn't matter what you called it: it was a random pattern of punctate pigmentation that matched the skin of one individual and not the other.

After a short deliberation by the jury, Ryal was found guilty of sexual assault on a minor and making indecent images of a child. Whether or not Gayle Ryal was subsequently charged with perjury I do not know, but as for her husband, he was sentenced to eighteen months in prison and placed on the sex offenders register for ten years. In prison he was considered unsuitable for sex offender treatment because he continued to maintain his innocence.

I remember sitting in the witness room when the young victim came in with her family. She had already given evidence at one trial, and she was utterly distraught, shaking, crying and insisting that she just couldn't go back into the courtroom again. It is hard for anyone, but especially someone so young, to have to relive their trauma in front of strangers, some of whom are intent on proving them to be a liar. It is no wonder that so many cases of rape or sexual assault are never reported or don't reach the courtroom.

But in those that do, forensic anthropology can help in many ways to bring the offenders to justice. I have often been asked if the work we do makes offenders more careful, and the answer is that it doesn't seem to. I firmly believe there is no part of the human anatomy that cannot be of some value to the identification of the victim, the prosecution of the guilty or the exoneration of the innocent. Our work is not restricted to what we can read in the bones. And as techniques and technologies continue to advance, so the evidence our bodies become capable of revealing will increase.

# 9

# The Hand

*"We find only one tool, neither created nor invented, but perfect: the hand of man"*

Julio Ramón Ribeyro
Writer, 1929–94

When I look at my hands, I see my father's very large and capable shovels, not, unfortunately, my mother's more delicate and feminine version. Probably best described as robust, they would never be chosen for a hand-cream commercial or to grace the front cover of *Vogue*. But they are mine, and they have done everything I have ever asked of them. They have dissected hundreds of human bodies, typed my thoughts, held my babies, wiped tears and bottoms and dug corpses out of a septic tank on a freezing cold day in January. Only now are these trusted and valued servants starting to complain, just with a twinge now and again.

The human hand is a miracle of evolution working in harmony with engineering. The American palaeontologist and biologist Stephen J. Gould was right when he admonished his fellow scientists for paying too much attention to changes in the human skull when analysing and presenting evidence for evolution. He believed they were looking in the wrong place: they should have been looking at the hands. Was it intelligence and big brains that allowed us to become such expert manipulators of our world, or was it when we stood up on two legs and freed our hands that we were finally able to challenge our brain to live up to its potential and keep pace with our dexterity?

Perhaps we only really understand the value of something when we lose it. Try to imagine your life without one or indeed both of your hands. You might think it is only those serving in combat zones around the world, where roadside bombs or improvised explosive devices are a constant threat, who need to worry about losing parts of their bodies. But we don't have to be on active service to put ourselves at risk: simply doing nothing is enough. Some six thousand upper and lower limb amputations occur each year on the NHS as a result of type 2 diabetes.

We use our hands not only for practical purposes but to clasp, hug, caress and to greet. The hand is served by a wealth of nerve endings that convey touch and warmth directly to the brain. The fact that almost a quarter of the body's sensory capacity is given over to the hand almost justifies elevating it to the status of an organ of sensory exploration in its own right.

Modern technological prostheses may be capable of replacing some of the motor functionality of the hand, enabling amputees to undertake basic tasks, but we do not have the technology that can recreate its innate sensitivity or the human connections made by a living hand, either for the toucher or the touchee. Even replicating the complex and subliminal gestures we use to communicate remains beyond the scope of replacement limbs, amazing though they may be in what they can now do compared with their more primitive predecessors.

Most of us find it extremely challenging to talk without the involvement of seemingly involuntary hand movements that either directly convey certain points or emphasize them. When we cannot use our voice, we have learned to rely on our hands and mouths to produce gestures and shapes to express ourselves. Technology now provides new options for those who lose their sight, but the art of reading Braille, which entails distinguishing between and interpreting the world of information contained in a pattern of little raised dots, is a perfect illustration of the value of the volume of nerve endings we have in our fingertips.

The intricacy of the anatomy of the hand, and the fact that it is so often seen in motion, make it a challenging part of the body for an artist to capture with accuracy. The detail reproduced by Leonardo da Vinci, marvellous though it is, pales in comparison with Albrecht Dürer's drawings of hands, in which you can almost touch the veins, the tendons and the wrinkled skin. In his old age, and in poor health, Henry Moore drew his own hands as a representation of the ageing body. "Hands can convey so much," he said. "They can beg or refuse, take or give, show content or anxiety. They can be young or old, beautiful or deformed."

To get an idea of what a miracle of engineering the hand is, let's take one simple movement, the action of picking up and holding a pen, and look at what our body has achieved to enable us to perform it.

First of all, it had to develop a pentadactyl (five-fingered) limb. Paired elevations, or limb buds, begin to form in the neck region of a human embryo at around 26 days of intrauterine life. By day 33, we have a recognizable hand plate at the end of the upper limb. At this point it looks a bit like a paddle as there are no separate fingers yet. Five days later, the edge of the paddle takes on a crenulated appearance as the cells in the interdigital spaces start to die and the tips of the fingers begin to emerge. The fingers gradually become more distinct as the cells between them continue to die off.

When these cells don't do their job all the way down to the pre-destined position, we end up with webbed fingers. Sometimes the fingers don't separate properly at all, a condition called syndactyly, whereby two or more fingers (or, more commonly, toes) remain fused together. This can be remedied by a relatively simple operation to separate them, resulting in two fully functioning digits.

By day 41, the neurovascular structures have penetrated deep into the hand plate, ensuring that future soft tissue will have access to a blood and nerve supply and all of the thirty-four (or so) muscles required to operate each hand will be in working order. By day 47, the hand can rotate, and by the next day cartilage, the precursor of the bones, will have started to form. Over the next eight days, cells die in pre-programmed locations in the cartilaginous mass to create joint spaces between the future bones. This is vital if the hand is to become a flexible multi-functioning tool and not remain little more than a static shovel.

By day 56, the thumb, known anatomically as the pollex, has rotated into a different plane from the other digits to produce our prehensile grip. The ability to bring the pad of the thumb into contact with the pads of all the other fingers is known as opposition. This is the feature that distinguishes us primates from the rest of the animal kingdom: your average dog, cat, horse or capybara simply can't do it.

By now swellings have also appeared on the pads at the end of the fingers, which are jam-packed with the nerve endings critical to the sensitivity of the hand as an exploratory tool. This is also where our fingerprints will form.

And so, providing all has gone well, six weeks into your development, you will already be equipped with a perfectly functioning pair of hands. You will quickly learn to draw comfort from them (very early on, babies can be seen on ultrasound imaging sucking their thumb or a finger). It will take you a little while after you are born to learn how to use them smoothly and with precision, but the "reach" and "grasp" reflexes can be strong in newborn babies. Some palaeontologists believe this to be a remnant of the need to cling on to our mothers tightly during a previous arboreal existence.

So, we have the required pair of hands. What do we need to do with them to pick that pen up? First, our brain recognizes it as a pen, and we have had the thought motivating us to pick it up. To reach out for it, our brain needs to send impulses from our precentral cortex, down our spinal cord and out through the spinal nerves that serve the upper limb (located in the neck region, where the limb first formed). These pass through the brachial plexus, a spaghetti junction of nerves in the armpit, and then fan out to the muscles we have decided we need to move to perform our action. We must flex the deltoid to lift our upper limb, the serratus anterior to propel it forward and then contract at least six muscles in our forearm to activate the wrist and the joints of the index finger and thumb. To make sure the movement is smooth, our cerebellum at the base of our brain oversees the operation and irons out any potentially ragged manoeuvres.

Now we can feel the pen between our thumb and index finger through those sensory nerves in our fingertips, which have sent their signal all the way back to the post-central region of the brain to tell it that the pen is in our hand, confirming what our eyes are also reporting. Although we think we are feeling the pen in the pincer action between the thumb and index finger, this is really taking place in the brain.

To hold the pen where we want it, we then turn our wrist to a semi-prone position, using a couple of muscles in our forearm, and arrange our thumb so that it is flexed at both joints and our index finger so that it is flexed at two joints but extended at the third. All the other digits of the hand adopt a relaxed but contracted position so that we can tuck them out of the way in our palm.

And all this has happened before we have even started to think about what we are going to write. The human body is utterly amazing and none of its accomplishments more so than the delicate ballet performed by the hand, all reliant on a vast supporting cast that have been rehearsing their respective roles since long before we were born. We make all of these movements without even thinking, just taking it for granted that everything will be in the right place, correctly wired.

Clearly such feats demand a complex underlying structure, and so it is not surprising that of the 200-plus bones in the adult human skeleton, at least fifty-four of them, over a quarter, are found in the paired hands. The bones are small as their versatility and flexibility of movement require short segments to accommodate muscle attachment. There are normally eight carpal bones in the wrist region, five metacarpals forming the flat of the hand and fourteen phalanges (three in each digit apart from the thumb, which only has two) with a couple of tiny sesamoid bones thrown in for good measure in the tendons of muscles associated with the thumb.

The small size of the bones makes them hard to recognize outside the context of the rest of the skeleton. This is certainly true of children's hands, as their components can be so tiny they could be mistaken for something like lentils, grains of rice or small stones. And it is not unusual for forensic anthropologists to have to search for hand bones, as each of the bony parts may fall away as the body decomposes. Human hands tend to be uncovered, and they stick out of the bottom of sleeves, making them easy prey for scavengers to carry off. So, as we have seen, when a body is discovered without its hands, we are not immediately concerned that they have been removed in a criminal act, although we do, of course, always check for cut marks on the remaining bones, just in case. In most circumstances what we will find are the marks left behind by the canine teeth of a fox, sometimes a badger. Feral cats and dogs will also target hands.

While it is fairly common for a body to be found minus its hands, what is less common is a hand turning up without a body. Of course, if one does, it doesn't necessarily mean that its owner is dead, as it could be the result of amputation, either accidental or deliberate. And fingers have been known to be removed from kidnap victims to extort ransoms, even if this is a rarer event in real life than it is in crime novels and films.

When an isolated hand or finger is found, how do we know that it is human? My colleagues and I are used to police officers phoning to tell us that they have found a hand on the beach. So used to it, in fact, that our initial response is inclined to be almost blasé. Before even looking at the images we will ask them to send us, we will probably offer the opinion that it is likely to be a seal flipper. It is amazing just how closely a decomposing seal flipper can resemble the human hand. A seal flipper is, like the human hand, a terminal appendage on a pentadactyl (five-digit) limb. There is some debate about the evolution of the pentadactyl limb, but it is fairly characteristic of all four-footed animals, which include amphibians, reptiles, birds and mammals.

It is likely that these appendages evolved from the paired fins of primitive fish as they adapted to the need to move around on land. The basic form has been modified in different species, mainly through the loss or fusion of bones in the "foot" or "hand." The ungulates are a good example of this. Their pentadactyl limb has evolved hooves to meet the specific requirements of their form of locomotion. Some are odd-toed (perissodactyl) ungulates, such as horses, others even-toed (artiodactyl) ungulates, such as camels; there is also an order "subungulates"—paenungulata, meaning "almost ungulates"—which includes elephants.

We had one of these routine "hand found on the beach" phone calls one day from a police officer on the west coast of Scotland. We went through the usual motions, asking him to send us a photo, which we would, of course, check, while informing him airily that it was most likely to be a seal flipper and he shouldn't be too worried. Such reassurances take the immediate pressure off the police. Investigating the provenance of a dismembered human hand would mean gearing up for a large-scale operation involving air, land and sea searches and coroners and procurators fiscal, not something it is advisable to launch prematurely.

However, when the photos arrived it was clear that this was definitely a hand, not a flipper—and it was very nearly human. But not quite. No visible skin remained owing to advanced decomposition but the proportions were all wrong: the thumb was very short and the fingers very long. This was the hand of a non-human primate, probably a chimpanzee. There was no evidence of cut marks to suggest it had been removed with a bladed implement, and no obvious signs of predation, either. How on earth do you get a chimp's hand on a Scottish beach?

Perhaps it was from a wildlife park, or a sanctuary specializing in rescuing non-human primates; perhaps from the burial of a pet. Or it could have been dumped overboard from a boat shipping illegal animal parts for homeopathic medicines or black magic practices. We never found out, but we never again assumed that every hand on the beach would be a flipper, and it made us think twice about nonchalantly voicing this opinion before taking a proper look.

◊

As we saw in Chapter 8, hands, and fingers in particular, are often lost by those who die in a fire. As there is very little soft tissue or fat covering the hands, it doesn't take long for them to burn down to the bone and for the bone to crumble into ashes. So it is important when recovering a body from a fire scene to thoroughly search the area around the base of the forearm to ensure that all the ashed fragments of the hand are retrieved. Given that these are so difficult to identify, it is increasingly being seen as essential to include a forensic anthropologist in any team dealing with a fatal fire, during both the recovery and subsequent investigation stages.

Our skills and anatomical knowledge have proved to be of genuine assistance to police officers and fire investigators. They are always amazed that we can pick up a tiny fragment of burned matter, which to them looks like nothing more than a piece of charred wood or a small stone, and tell them that it is a finger or a wrist bone.

Unfortunately, the input of forensic anthropology has not always been seen as crucial in these investigations. It is the way of the world that, too often, it takes a mistake to bring about improvements in procedures. This is precisely what occurred in one tragic case, in which the police only sought my opinion in the first place because the local forensic pathologist happened to be on holiday.

There had been a fire that, heartbreakingly, had resulted in the death of two little boys. It was sparked by an electrical fault in a remote, idyllic Victorian cottage in the Highlands which went up like a tinder box because of its original pitch pine features. The fire engines had a long way to travel and had to negotiate a winding, single-track road to finally reach the cottage. With nobody to help until they arrived, the parents battled the flames to try to rescue their sons, who were trapped in their bedroom, but they were beaten back by the ferocity of the fire and the thick, black smoke. It was believed that, mercifully, the boys probably died in their beds as they slept, overcome by the smoke. I cannot imagine the torment of watching your house burn, with your children inside it, powerless to save them.

Eventually the fire service got the blaze under control and, when the building was finally declared safe, they began the grim search for the bodies of the two boys. The roof had collapsed and the rafters and slates had caved into the cottage, so everything had to be lifted out of the way by hand as the shell of the building was searched, room by room. The heavy rafters were removed carefully and stacked outside at the front of the house. The boys, still in their beds, were found buried under slates and timber. Once the debris was cleared, their badly burned bodies were transferred to the mortuary for examination.

One child had been found intact but it was clear that significant parts of the older boy were missing. It was explained to the family that it was likely the rest of the body had been completely incinerated and would not be recovered. The children were buried in little white coffins as the mourners marvelled at the stoicism and dignity of these parents who had lost everything.

The couple returned to their cottage regularly as they tried to come to terms with their loss and work out what could be salvaged from the ruins of their home, perhaps as mementos of happier times. Two weeks after the fire, they were there laying flowers, as they always did, when they noticed a little pile of bones on the grass in their garden.

They contacted the police and were reassured by an officer who came out to have a look at the bones that they were most likely to be from an animal. He was a local, a countryman, and he thought they were probably cat bones. He told the couple not to worry, scooped the bones into an evidence bag and took them to the mortuary. As the forensic pathologist was on holiday, the police asked me if I could please go over to the mortuary and give these cat remains the once-over to reassure the family that these were indeed what they were.

When there is an anticipation that this kind of examination is a formality and you will just be going through the motions, nobody is really interested in what you are doing, so support in the mortuary was minimal. Unfortunately, this was to be one of those occasions when I was going to rock the boat, because I was about to utter the unimaginable: these little "cat" bones were, without a shadow of a doubt, human, and from a young child of between four and six years of age.

There was an assortment of bones, including parts of the vertebral column, little fragments of rib and some of the small bones from the wrist. Several of them bore tooth marks. To the police, they were just little ivory-coloured flecks that could have been anything. The easiest course of action in such a situation is to challenge the anthropologist, especially since I was telling them something they were not happy to hear.

Asked if I was sure about this, I confirmed that I was. Nevertheless I was questioned hard about how I could be so certain. I replied that I had written the textbook on the identification of juvenile bones. Not only could I tell them which region of the body the bones were from, I could tell them which side, name them all individually and give them an age estimate.

The temperature in the mortuary dropped another ten degrees. It was, it seemed, more acceptable for the expert to be wrong than for an unwelcome opinion to be correct. As the police left me to finish my note-taking, I asked if a report was required from me. Surprisingly, I was told that it was not. They would rather wait for the forensic pathologist to come back from his holiday.

I went home. I felt very uncomfortable that something I knew to be categorically true was being questioned and that there was nothing I could do about it. I thought about the child, the family and the fire. I thought about how the bones might have got out into the garden. I thought about the implications of what I had to say going unsaid and I decided to write a report anyway, if only for my own peace of mind. Experience has taught me that, however much you think you will remember, if you don't write it down at the time, you quickly forget details. I also know that if something is not recorded in writing there will be no evidence that it ever took place.

A couple of weeks later, quite by coincidence, I was contacted by the family's solicitor. It appeared that the pathologist had returned from his holiday, looked at the bones and confirmed that I was right, and the police had contacted the bereaved couple to inform them that the remains they had found in their garden did in fact match some of the missing parts of their elder son. The parents were seeking a second opinion and their lawyer, someone I happened to know, wanted to retain my services. Imagine his surprise when I told him I had already written a report on the case.

He was unaware that I had examined the bones because, it seemed, my presence in the mortuary had been airbrushed out of the police records. This was possible because there was no report. If I had been asked to submit one, my examination would have had to be disclosed. I returned to the mortuary, this time on behalf of the family, to reassure myself that the remains I had examined were the ones being returned to them. The bones were all in agreement, they were human, from a young child and some showed evidence of predation. Everything was present and correct, and I gave my original report to the lawyer, together with an addendum covering my second visit to the mortuary.

And that was the end of it. Until, some time later, I had a call from another police force in the south of the country. Apparently, they had been brought in to investigate the handling of the case by the original force to try to establish what had gone wrong and what lessons could be learned. The SIO suggested that, rather than visiting me in my office, perhaps he could come and talk to me at my house, as he lived not far away and it would be more relaxed and less formal. He arrived with a colleague, and we all sat in my kitchen, drinking coffee and eating biscuits through what was still a long interrogation, less formal or not. The SIO asked the questions while the junior officer recorded my every word.

I could not say why I had not been asked to write a report: they would have to ask the local police about that one.

Why had I written one anyway? Because my professional advice had been sought and I had carried out the work requested. Whether or not I was going to be paid to produce a report, it was my duty to record what I had found.

How did I know at the time that the remains were human? At least by now the bones had been tested for DNA and it had been confirmed that they belonged to the little boy. So there was nobody doubting my ability to identify juvenile remains, just someone trying to understand why others would have chosen not to accept it.

How had the lawyer happened to come to me for advice? The answer to that one was easy: pure chance. Although he knew me both professionally and personally, he had no idea when he approached me that I had already been involved.

Then the question I'd been waiting for. How did I think the child's remains had come to be found, two weeks after the fire, in the middle of the garden? Of course, I could not say. I had never visited the scene and I had not been present during either the original recovery of the bodies or the subsequent discovery of the bones.

All I could do was surmise, and I asked the SIO if he wanted me to do that. He did. I had been going over the puzzle in my head and I had come up with a theory. It may be completely wide of the mark, it may be partially correct or it may be true. We will never know. But it is at least capable of plausibly explaining what happened.

During the fire, the roof had collapsed into the boys' bedroom. They had been buried under the debris, and it was known that wooden rafters had fallen across the children's beds, so it was possible there had been contact between the body of the elder boy and the underside of a burning rafter.

A burning piece of wood will sear and adhere to human skin, and in doing so it can protect that part of the body to some extent from further damage. Provided the wood is not totally consumed by the fire, body tissue may remain stuck to it even when it is finally lifted.

Once the flames had been extinguished and the fire service were clearing the rubble looking for the bodies, they painstakingly removed the collapsed rafters from the bedroom. Perhaps they never turned them over to inspect the undersides, and therefore wouldn't have noticed small parts of a child's body stuck fast to one of them. The rafters were then stacked outside the house along with all the other debris.

Here the decomposing tissue would have been detected by animals. Cats and foxes in particular have an excellent sense of smell. They would have searched out the remains and carried them away for consumption—further out into the garden, where the bones had been found. The tooth marks I saw on the bones supported this supposition.

The SIO queried why I had not tendered this as a possibility to the police at the time. The answer was simple: nobody had asked.

Police forces have long memories and I didn't work for that particular force for at least a decade. This fire took place thirty years ago, and times and police procedures have changed a lot since then, which can only be a good thing. Even though the role of forensic anthropology in fatal house fires is now widely recognized, sometimes it is still a struggle to have our skills acknowledged.

When we recover an adult body that has been fragmented by a fire, we know we are looking for twenty-seven hand bones, or remnants of them, in total—8:5:14 is the magic formula: eight carpals, five metacarpals and fourteen phalanges. However, not every hand has that exact complement, and while we expect to have four fingers and a thumb, in all the right places and of the right proportions, this isn't always the case. There may be variations in the structure of a hand caused by congenital or accidental alterations.

As we have already discussed, much of the future appearance of the hand is set between four and six weeks of fetal growth. Anything that interrupts normal development at this time may be manifested in the final appearance of the hand, although genetics will also play a significant role in human variations. Probably the most common congenital conditions are seen in the number of digits present on the hands.

The scientific name for the arrangement of fingers on our hands, or toes on our feet, is dactyly. Having too many is called polydactyly, which is most frequently represented on the hand by an extra vestigial digit on the medial side of the small finger. It is a relatively simple surgical procedure just to snip this off. Usually the digit is comprised simply of soft tissue, although sometimes a full extra bone will have formed. Polydactyly is the result of a non-vital genetic mutation and so can be passed down through the generations of a family. It is not uncommon, affecting about 1 in every 1,000 births. In 2016 a woman in China exhibiting polydactyly (she had 6 fingers on each hand) gave birth to a son with a more extreme form of the condition: he had 2 palmar regions for each limb, 7 fingers on his right hand, 8 on his left—15 fingers in all, but no thumbs. He also had 8 toes on each foot. An incredible total of 31 digits, 11 of them supernumerary. But even this is not the world record. The highest number of digits ever recorded was 34, belonging to a young Indian boy born in 2010 with 10 toes on each foot and 7 fingers on each hand. As he later had some of these removed, the official world record for living with polydactyly is held by another Indian, Devendra Suthar, who has 28 digits, 7 on each hand and foot. He is a carpenter and says he has to take extra care when cutting wood not to chop any of them off.

Oligodactyly is the opposite condition: fewer digits than we would normally expect. This is usually associated with a range of clinical syndromes. Ectrodactyly, or split hand/split foot formation (SHFM), is the absence of one or more central digits. This reduces the hand or foot to an even number (either four or two), giving it the appearance of a claw. In those who have only two visible digits, it is likely that syndactyly, fusion of digits, will also have occurred.

Macrodactyly, which is rare, is when the fingers or toes grow to an abnormally large size. This tends to be seen only on one hand, most commonly in the index finger. The cause is not well understood. Conversely, brachydactyly results in very small digits, usually because of shortened bones. An inherited condition, brachydactyly is generally present at birth but only becomes noticeable as some of the fingers start to grow and others do not follow suit.

There are variations on these conditions that are most unusual, among them mid-ray duplication polydactyly, which produces a duplicate finger. Thomas Harris gave his character Hannibal Lecter a duplicate middle finger on his left hand in his novel *The Silence of the Lambs*, although this singular feature did not make it into the film version. Equally rare is a transposed finger. I was giving a lecture in a pub one evening (as you do . . . something called a Pint of Science) in which I was talking about my research into hand identification. A young woman asked me afterwards if I'd like to take a picture of her hands as she had been born with her middle and ring fingers transposed. And recently another lady let me photograph her hands, which had an extra transverse crease on the little fingers that didn't align with a joint. She considered it weird; I thought it was really interesting as it is something that occurs in less than 1 per cent of the population. I love hands. They are so quirky.

Other alterations to the hands, of course, may occur during the life of an individual. Amputations, due to accident, ritual, surgical removal or, in some parts of the world, punishment for a crime, are a common anomaly. Self-amputation, or autotomy, is seldom seen, and is generally the result of a person becoming trapped and having to remove their own limb to free themselves and save their own life. It is also reported in people with body identity disorder, where the sufferer cannot recognize parts of their own body as belonging to them and feels compelled to remove what they perceive as an offensive imposter.

Amputations can also occur in utero, caused by the presence of amniotic constricting bands. This, too, is very rare, but quite a shock for the new parents.

Forensic anthropologists, then, need to be alert to the possible manifestations of any one of these conditions or events. In cases of amputation we will always look at the ends of the bones to see whether there has been any healing to indicate that it occurred before the person died. Cut ends would point to perimortem or postmortem amputation, and we will usually be able to detect the marks of the tool used to remove the body part.

In the days before attention was paid to the health and safety of manual workers, missing fingers were an occupational hazard. Recently, some nine hundred historical pictures were exhibited of convicts, about a third of them women, photographed just before their release from HM General Prison in Perth, Scotland. These were basically nineteenth-century police mugshots, taken to help the enforcers of law and order to keep track of criminals as they moved around Scotland. There was the usual full-frontal facial view, with a profile captured by a carefully positioned mirror. But what was really interesting was that many of them also had their hands photographed. Apparently, this was to record that they were still in possession of all their fingers—or, in some cases, not—at the time of their release as amputations in industrial accidents were so common that the loss of a digit became as much of an identifying feature as a face.

These days there is a vogue for "amputation tattoos" that acknowledge the absence of a finger, sometimes an entire hand. For example, the phrase "Good uck" spelled out across the knuckles, with no "L" because the finger where it should go has been amputated. I have also seen "Love" tattooed on the knuckles of a man's right hand, while the back of his left hand, from which all of the fingers had been amputated, carried the message "No room for hate."

We are all used to practical jokes involving fake severed fingers, especially around Hallowe'en. When I was a child, every October one would turn up in someone's serving of school rice pudding. These days, rubber fingers are not, it seems, enough for some. Grotesque gothic jewellery made from archaeological bone is now bought and sold on the internet. When you look at the online comments of those prepared to spend $15 on a necklace made from real human finger bones, it pretty much says everything you need to know about this kind of person.

"Bones were clean. One showed osteoporosis or arthritis. Great gift for my friend."

Or this, from the Q&A section: "Q. Are the three bones from the same finger?"

"A. The bones are from separate people, but we try our best to select those that look like they might go together."

Does anyone else wonder on what planet it is OK to buy bits of dead people and sell them as jewellery, or is it just me?

I was working in my office one day when Viv, my PA, put through another of those phone calls from the police.

"We have something unusual we would like you to look at. Can we bring it over?"

This time it was clearly not going to be a seal flipper. Two officers arrived in the office with a small evidence bag. This being Scotland, they always hunt in pairs and accept a cup of tea before we get down to business.

Inside the bag was a silver-coloured key fob which had been found in the undergrowth at the side of a wooded path by a man walking his dog (it is always someone walking their dog). There were no keys on it. Nothing strange about that, until you looked at what was dangling from the key ring: three human finger bones, perfectly articulated, strung on a silver wire that bridged the joint spaces. I looked at the policemen and they looked at me. Their eyes were pleading with me to tell them it was just a piece of novelty tat and I was about to ruin their day.

These were the distal, middle and part of the proximal phalanx of the left index finger of a young adult male. They had been cleaned (probably boiled and bleached) and there was no detectable smell of decomposition about them, so they had probably been separated from the rest of the hand for some time. The cut marks on the proximal end of the proximal phalanx indicated that the finger had been removed by a saw, probably an electric one rather than a handsaw, judging by the spacing and regularity of the cut marks.

Now the police had to investigate. They began by making door-to-door inquiries in the area where the item had been found. Given that we knew we were looking for a young male (most likely still alive) who had lost his left index finger, if he did live locally he wasn't going to be too hard to find. Sure enough, the police soon tracked him down.

The key fob belonged to the amputee, David, who had worked as a carpenter in his father's business since he was a boy. One day, needing to cut some wood in a hurry, he had bypassed the required safety guards and protection and, in his haste, had an accident with his circular saw and chopped his finger clean off. He and his finger were rushed to hospital. The finger could not be reattached but he asked if he could keep it and was allowed to do so.

There are regulations governing what happens to amputated body parts and most of them are, of course, incinerated as medical waste. But there can be a bit of wriggle room when patients ask to keep bits of themselves such as gallstones and teeth. While individual hospitals have their own policies, there is no law against people retaining their own body tissue, as long as it doesn't constitute a public health hazard, and amputees can and do ask for assorted body parts to be returned to them. The view of the Human Tissue Authority is that when this is permitted, hospital records should be kept to ensure traceability.

Some amputees want their whole body to be reunited in death, for religious or personal reasons, and to save whatever they have had removed to be buried or cremated with them when they eventually die. Historically, the limbs of people who lived on without them were sometimes given their own graves. One such was the leg of Lord Uxbridge, shattered by a cannon shot in the Battle of Waterloo. It is he who is supposed to have coined the expression "one foot in the grave." Recently this custom has made a comeback, with a Muslim hospital chaplain in the north of England setting up a public burial site specifically for amputated limbs.

One thing you cannot do with your own amputated body parts in your lifetime, ironically, is have them cremated, as the Cremations Act 2008 does not allow human tissue from living people to be accepted. And yet there is nothing to stop people doing it themselves on a bonfire.

Others go for some outlandish options, often after having their amputated limbs rejected for scientific research. One woman in the USA apparently has an amputated foot, which she paid to have skeletonized, with its own Instagram account. Words fail me.

But ours is not to reason why, and it may be argued that, inexplicable as it may seem for an amputee to want to keep a part removed from their body, they ought to be entitled to do whatever best helps them to deal with their trauma. Should we all have rights over our own bodies, attached or detached? That is up for discussion, but when it comes to the lack of respect shown for bits of other people's, as demonstrated by the distasteful trade in trinkets made from human bone, I believe that a line should be drawn in the sand.

Still, what David did with his amputated finger is mind-boggling by any standards. He took it home and boiled it in a pan of hot water until all the soft tissue fell away. He said he had seen his mum doing this with bones when she was making soup, so he thought it would probably work. The soft tissue and fingernail went into the bin and he popped his finger bones into a container of bleach until they were nice and white. But when he noticed that they were still leaking a little fat he decided to boil them again, this time using a biological detergent, which was what his mum did when she wanted to get fat stains out of the tablecloth. I promise you, this is all true. He then put the bones on paper towels on his bedroom windowsill to dry in the sun, and when they stopped smelling and leaking, he transferred them to a small glass jar, which he stored on a bookshelf in his bedroom.

He wanted to keep them because he thought they were "cool," but for a while he didn't really know what he was going to do with them. Occasionally he would take them down to the local pub to show them off to his mates, especially at Hallowe'en. He often remarked that maybe one day he might make them into jewellery, and eventually he decided that this was what he would do.

He drilled a hole lengthwise through each finger bone and threaded them together with silver wire, which he knotted at the end of the distal phalanx, made a silver cap to cover the cut end of the proximal phalanx and mounted them on a key ring. As if this were not bizarre enough, he then chose, unaccountably, to present the key ring to his new girlfriend on Valentine's Day as a token of his undying affection.

She may have been hoping for roses or chocolates, perhaps even a diamond ring for her own finger. Whatever the case, she was so disgusted that she threw the key fob into the bushes—much to David's consternation, as he was unable to find it, despite searching for hours. Suffice it to say that the relationship did not survive. He hadn't broken the law, so perhaps his only crime was really bad taste in gifts. This may sound like one of those shaggy dog stories that is embellished with every telling, but it came to us direct from the police officers who tracked down our bold, nine-fingered Romeo.

How did we know that the finger bones originated from a young, though adult, male? Once again, it is down to the development of the bones. We have seen how the hand is already formed in the fetus: at birth, it will have nineteen identifiable bones, and at two months, bone will have started to appear in the carpals. The last one, the pisiform (pea-shaped) bone, is formed by around eight years of age in girls and ten in boys. Over the next seven years the hand will continue to grow until the growth caps at the end of each bone eventually fuse and the hand ceases to increase in size.

David, our unfortunate carpenter, lost his finger when he was sixteen. We could see on the X-ray of the bones from his key ring that the bases of both his middle and distal phalanges had fused to their respective shafts, but the fusion was not quite complete as "ghost" growth lines were still visible at the fusion sites. So we knew this had happened relatively recently. The bones of the hand are subjected to fluctuating levels of blood steroids throughout puberty and respond accordingly. Testosterone produces bones that are more robust and larger, and because a boy will be two years older than a girl by the time his hand stops growing, the effect of this hormone in particular results in bigger bones than would be the norm in a female hand. This enabled us to offer the opinion that the bones belonged to a male.

However, the X-ray of David's finger could only tell us how old he was when it was chopped off, not how long ago this had happened. His finger, in its glass jar and latterly attached to its key ring, remained for ever frozen at sweet sixteen, whereas he was now eight years older— though it would seem not much wiser—by the time the police felt his collar.

Because there are so many bones in the hand and they are all growing and maturing at their own rate, a radiograph of the hand is often used as a means of determining the living age of a young individual.

When an undocumented refugee or asylum-seeker enters the UK as an unaccompanied minor (UM), their likely age will need to be assessed. Many do not know exactly how old they are because the country they have come from does not keep such records. Others may have fled without their documentation, or lost it en route, and so have no definitive proof that they are the age they claim to be.

Social workers will usually assign to these children what they consider to be their most likely age, based on their responses to various questions, on how old they look and on their general maturity. If the individual is deemed to be under eighteen years of age, they will be assigned a new birth date of 1 January in the most appropriate year. Deciding as far as possible whether somebody should be categorized as a child or an adult is important for the purposes of child protection. As the UK is a signatory to the UN Convention on the Rights of the Child, if we believe an individual to be a child, defined under the treaty as anyone below the age of eighteen, then we must, among other things, house, educate, feed, protect and care for them until they become an adult. The child will then be the responsibility of the local authority until they reach the age of majority.

Like all systems, this is open to manipulation by the unscrupulous, in this instance, adults who seek to take advantage of looking younger than their years to benefit from rights granted to children, and who may deliberately travel to the UK without their papers. If they are assigned an approximate birth date below the age of majority, the truth may never come to light provided they keep a low profile and don't break the law.

Our justice system has its own criteria involving a number of age thresholds. A child under ten, for example, is deemed to be below the age of criminal responsibility and juveniles (children between the ages of ten and seventeen) who commit an offence are subject to different procedures in dealings with the police and the courts from adults. Although anyone over eighteen is considered in the eyes of the law to be an adult, twenty-one is the minimum age at which you can be sent to prison. If you are under twenty-one and given a custodial sentence, you will therefore serve it in a young offenders' institution.

In cases where there is any doubt over an offender's age that has a bearing on how they are categorized by the criminal justice system, the courts understandably require more definitive evidence than an age that has simply been assigned by social services, and so it is usually when young refugees or asylum-seekers fall foul of the law that forensic anthropology becomes involved.

Majid was a refugee from Afghanistan who arrived in the UK as an UM and was assigned an age of sixteen by social workers. He was placed in the care of a local authority home for the remaining two years of his childhood. It was here that he started to groom young girls. Two years after he left the children's home, he was arrested for the rape and murder of his girlfriend's best friend and found himself before the courts.

Majid's records showed that he was now twenty years old. However, his girlfriend told investigators that he had said he was actually twenty-four, and had boasted about having fooled the authorities. If he was twenty-four, then it would mean, of course, that he had been twenty when he moved into the children's home, not sixteen, and that vulnerable children in care had been exposed to a predatory adult masquerading as a child.

Naturally, the court needed to know this man's true age, or the nearest we could get to it, so we were asked to examine him. Lucina led on the case. First of all, radiographs were taken of his hands, which clearly showed that all the bones had fully fused—proof that by this point he was certainly older than seventeen. A CT scan of his clavicle indicated that he was closer to twenty-five.

As a result, when Majid was duly pronounced guilty, he found himself serving his custodial sentence in prison, not in a young offenders' facility. He had not been a child even when he first entered the country, let alone when he began offending. He had arrived as an adult. He will therefore be deported on his release.

This case is an example, albeit an extreme one, of why it is crucial that age assessment in the living is undertaken scientifically. It is too important, not only in protecting the rights of the person being assessed, but also the rights of others, to be left to guesswork. We have the expertise to do this, thanks in large part to the story the bones of the hand can tell us. I am of the firm opinion that the procedures used in age determination of the living are in need of a thorough overhaul. Clinical imaging of the hand is a reliable indicator of age and perhaps it should be used more routinely. Concerns that X-ray radiation may be harmful are easily resolved: it is perfectly possible to use MRI images instead as these involve no ionizing radiation. If Majid's hand bones had been scanned when he first entered the UK, the authorities would have been in no doubt that he was not telling the truth.

The hand is, of course, of huge value in identification. It offers an array of information, from the arborescent pattern of our superficial veins and the pattern of skin creasing over our knuckles to the location, orientation, size and shape of our scars and the number and distribution of our freckles, moles or liver spots. And it is the hand, or rather the print it leaves behind, that is the location of a biometric that has for centuries been accepted as confirmation of our identities. Fingerprints have been found in ancient clay tablets, representing the signature of the person who made them, and were used by Chinese merchants to seal contracts.

The patterns seen in fingerprints were first described by the Italian anatomist Marcello Malpighi around 1686 and it was a German anatomist, Johann Christoph Andreas Mayer, who observed, nearly a century later, that they may be unique to each individual. In the nineteenth century, the Scottish doctor and missionary Henry Faulds published a paper that suggested fingerprints could be used in the investigation of crime. The baton was quickly picked up by the explorer and anthropologist Francis Galton, who published his seminal text on fingerprint identification in 1892. And the rest is, as they say, pretty much history.

We all know from our school biology lessons that every palm print and fingerprint is different, even those of identical twins, and each is believed to be unique. However, as this is impossible to prove, for evidential purposes we have to express the proposition that any two prints may have come from the same person, or from two different people, as a statistical probability rather than an absolute certainty. The world had almost reached the point of assuming that fingerprint identification was invincible until, in 1997, DC Shirley McKie, a serving Scottish police officer, was accused of being present in a house where a murder had occurred.

Because all police officers and scientific experts have their fingerprints and DNA recorded for exclusion purposes, DC McKie's were on file. When a thumbprint found on the bathroom doorframe of the house was run through the system, it produced a match with hers. She denied having been there, was suspended, then sacked, subsequently arrested, tried and eventually found not guilty of all charges because she had not been at the crime scene.

Amid allegations of misconduct on the part of the Scottish Criminal Records Office and the police, what became known as the

Fingerprint Inquiry was set up by the Scottish government in 2008. This shook the world of forensic identification into the recognition that while the prints themselves may indeed be unique, the methodology used to compare them may sometimes be insufficiently robust. The inquiry's report warned that, while there is no reason to suggest that fingerprint comparison is inherently unreliable, practitioners and fact-finders must give due consideration to its limitations.

It was a sobering reminder that every identification technique is fallible. Identification is not a matter of certainty, but of probability. Which is why all scientists must be able to understand the principles that underpin statistics.

The hand is also an area we examine closely when we are trying to establish how a person has died because it is so often the first part of the body employed to fend off an attack. The presence of defensive wounds may raise suspicions about what has happened to them, as they did in the case of a woman whose identity remained a mystery for some years.

Her semi-naked body, discovered by hillwalkers, had been found lying face down in a stream in an isolated spot in the Yorkshire dales. She was wearing jeans and socks and her bra was undone but still hooked over one arm. A T-shirt turned up some distance away, but there was no sign of any shoes, handbag or other possessions. She had been dead for no more than a week or two and the cold, flowing water had slowed decomposition.

After a postmortem examination failed to pinpoint any obvious cause of death, the body was frozen to halt any further decomposition or insect activity while the police continued their investigations. Some weeks later, they sought help from forensic anthropology. A second PM was requested and we made the journey south to see whether the body had anything more to tell us. We might turn up something that had been overlooked, or incorrectly recorded; if not, we would at least be able to confirm for the police the findings of the first examination.

The woman was between twenty-five and thirty-five years of age and about 4 ft 11 ins (1.5 m) in height. Her facial appearance, the colour and type of her hair and her dentition told us that she was likely to be of south-east Asian ancestral origin (perhaps South Korea, Taiwan, Vietnam, Cambodia, Malaysia, Thailand, the Philippines or Indonesia). Her jeans were a size 12, her T-shirt a 10. Her shoe size was probably about a UK size 1 or 2. She had a piercing in each earlobe and a wedding ring on her left hand made of gold that may have come from south-east Asia.

We found only two injuries, and they were both to her right hand. The first was a spiral fracture of her fifth metacarpal. The metacarpals are among the most common bones to be broken and constitute between 5 and 10 per cent of emergency department visits, most of these by young men. They are usually the result of a fall, a road accident, blunt trauma or assault—in either the batterer or the battered. Fractures to the head and neck of the bone are quite often caused by throwing a punch, whereas fractures to the base of the metacarpal, which are rarer, generally arise from high-force impact. It can be impossible to tell whether a fracture is due to something hitting the hand or the hand hitting something.

The second trauma was a dislocation of the woman's proximal interphalangeal (PIP) joint of the right middle finger, which may have been sustained in the same incident that was responsible for the damage to her little finger. This, too, could be explained by a fall, but it could just as easily have been a defensive injury.

No person matching this woman's description had been listed as missing and her DNA and fingerprints did not match anyone in criminal records. Isotope analysis of her bones, which gave us information about her diet, confirmed that she was likely to have been living close to the place where she had been found, and had been doing so for quite some time. Her face was drawn by a forensic artist and published in the local papers, but still nobody came forward.

Eventually she was buried in a small, rural cemetery as an unknown person. The money to pay for her burial was raised in the local community, who took care of her as one of their own until such day as she could be reunited with her name and with those who loved her. In the meantime, she was described on her gravestone as "the Lady of the Hills."

When a body remains unidentified, investigating the death is extremely difficult because the police have no means of tracing the last movements of the deceased. They cannot check their bank accounts, mobile phone records or computer or locate relatives, friends or colleagues for questioning. But even after all leads have dried up, cases like these are not forgotten. Cold case reviews take place on a regular basis. With the likelihood of solving them heavily reliant on scientific analysis, discussion is often focused on recent scientific or technological developments that were not available to the original investigation and which might now open up a new avenue of inquiry.

When this young woman was found, early in the new millennium, social media were in their infancy. With the expansion of these networks around the world, investigating officers thought it was worth trying to flood social media in south-east Asia with her description and the forensic artist's facial depiction to see if that would produce any fresh leads. And, quite incredibly, it did. Fifteen years after her death, the UK police were contacted by someone from Thailand who believed the Lady of the Hills was their relative.

A new lead like this ignites a cold case. Now that the police had a name to work with, they assigned a new investigative team, who flew out to Thailand to talk to possible relatives, collect familial DNA and obtain fingerprints from the local authorities who had issued the woman's ID card. Both DNA and fingerprints were a match. At last she could be named.

Lamduan had moved to the UK to marry an English teacher. As well as the two children they had together, she had an older son by a different father who had come to England with her. Shortly before her body was found, he had come looking for his mother and was told by his stepfather that she had left him and the two younger children and gone back to Thailand.

The couple had always kept themselves to themselves and had few friends. Lamduan's family in Thailand had lost contact with her around the time of her death, but as they, too, had been told that she had abandoned her husband and children, and thought badly of her for it, this was not entirely surprising. It meant her disappearance failed to raise alarm bells with them. Everyone simply believed the husband's story and assumed that she had run off with someone else, which meant that nobody, in England or Thailand, reported her missing or raised any suspicions.

Lamduan's death remains unexplained and the case is still open. All we have to go on anatomically are those two injuries to her right hand. And yet somehow she ended up dead, face down in a moorland stream, only partially clothed and without shoes or handbag. Could she have got those injuries in a fall, or were they defensive? If so, was she pushed? Who was there with her? We must not give up hope of one day being able to answer those questions.

How we adorn our hands can be a help in taking an inquiry in a particular direction when we need to identify a body, as demonstrated by Lamduan's wedding band, and the Claddagh ring found on the Irish lady in the case discussed in Chapter 2. So we always check the hands for jewellery, or a sign that jewellery has been present.

The hands are also a common site for tattoos; less so for piercings, although these are starting to become fashionable. They are usually web piercings, with a stud inserted between the thumb and the index finger, or between any two adjacent fingers, but we sometimes see dermal or single-point piercings on the wrist, on any finger or indeed anywhere on the hand.

In the future forensic anthropologists may have many more items on their checklists. There are reports that microchips containing personal information are now being embedded in hands to enable people to go about their daily business without the need to rummage around for an ID card, bank card or card key to gain access to their workplace. Some even hold information on their health.

One day we might not even need to carry passports as our entire identity could be implanted in our hands, or indeed any part of our body. Such technology might sound the death knell for some aspects of the forensic anthropologist's work. But not in my lifetime.

# 10

# The Foot

*"It's a fact the whole world knows, that Pobbles are happier without their toes"*

Edward Lear

Poet, 1812–88

I have always hated feet, living or dead. I hated dissecting them and I hated having to try to identify all those misshapen little nodular excresences that make up our bony toes. Feet have bunions, corns, callouses, warts, verrucas, gout. They can produce up to half a pint of sweat a day and they even make their own cheese. I hate it when you have to perform a postmortem on a decomposed body and you know that when you turn a sock inside out you are going to have to sift through the gloop of yellow-brown, slimy mush to find the lumps of bone. You might well find the floating toenails in this foot soup, and that sends a shiver down your spine. Gnarled, misshapen, fungus-ridden, thick slabs that have the cheek to ingrow: I hate them most of all.

In truth, feet are often overlooked during forensic postmortem examinations, which is ironic when you consider that they play such a big part in the iconography of the fictitious *CSI* forensic world, poking out cheekily from under a white sheet, usually sporting a fetching toe tag. It is also a mistake, because they do keep a lot of information hidden away under their arches. And that draws a glimmer of grudging respect from me.

To appreciate the foot we need to understand its purpose. The modern foot has two principal functions: to support the weight of our body when we are standing upright and to act as a mechanism of propulsion when we want to move. Pretty much nothing else.

The early twentieth-century naturalist and anatomist Frederic Wood Jones waxed lyrical about the foot: "Man's foot is all his own. It is unlike any other foot. It is the most distinctly human part of his anatomical make-up. It is a human specialization and whether he be proud of it or not, it is his hallmark and as long as Man has been Man and so long as he remains Man, it is by his feet that he will be known from all other members of the animal kingdom."

He was right: there is no other foot in the animal kingdom that looks like ours, and that is why palaeontologists get so excited when prehistoric human foot bones are found. A fossilized example from the Hadar region of Ethiopia showed that by about 3.2 million years ago our human ancestors were bipedal and walking on a modern-looking foot, a discovery that is supported by a number of other finds, the most important being the foot bone of a member of the *Australopithecus* genus of hominin from which we are believed to be descended: *Australopithecus afarnesis* AL 333-160. This specimen is a left fourth metatarsal and it is arched—a feature that is unique to the modern human.

In the human embryo, the lower limb starts to form around 28 days after fertilization, a couple of days after the upper limb has begun to develop. By day 37, a footplate resembling a paddle appears at the end of the limb and within four more days, digits are visible. Bones will begin to form towards the end of the second month. At birth, nineteen of the bones in the front and middle parts of the foot will be formed, plus the calcaneus, our heel bone, and the talus, which sits on top of it to make our ankle. Once growth is completed, each adult foot will have around twenty-six bones in total.

The calcaneus is the first foot bone to be visible on an X-ray, between the fifth and sixth month of our gestation, and the talus can be seen by the sixth or seventh month. The cuboid, the most lateral of the tarsal bones, may show bone formation just before we are born or within the first couple of months afterwards. In the past, looking at the developmental stage of these three bones was the most straightforward way of assessing the age of a fetus, and it was used by early pathologists to establish whether a deceased baby who had been delivered prematurely or aborted would have survived without medical assistance. Nowadays, of course, babies are viable from a much earlier age, but this information was often relied upon in the past to decide whether legal action should be taken against a mother.

Like our feet, our footprints are unquestionably human and there is no other animal that makes a similar print. Some or all of the heel, the lateral (outer) border of the foot, the ball and the pads of the toes may be visible in the impressions or marks we leave behind when our bare feet come into contact with a substrate, depending on the nature of the surface or material on which we have trodden. The medial (inner) edge will not leave a print as the internal structure of the foot raises this region into a series of arches that give our feet their elasticity and stability—the hallmark of the human foot.

Because a baby's foot leaves a fuller print, there is a popular belief that the arches do not develop until about two years of age. In fact they start to form quite early: it is the presence of a pad of soft tissue that gives a young child's foot its flatter appearance.

Ancient footprints preserved through time have helped to confirm the earliest dates established by archaeologists and palaeontologists for habitual human two-legged propulsion. It was another trio of Australopithecines who left behind some of the most remarkable evidence of their stroll across our planet's surface millions of years ago. The Laetoli footprints in Tanzania, a trail of about seventy impressions made in volcanic ash, were covered over by a further volcanic eruption and remained hidden for 3.6 million years until they were found by the celebrated British palaeoanthropologist Mary Leakey in 1976.

The Australopithecines walked in a modern heel-strike, toe-off mode, with a short stride that suggests a more diminutive stature than that of the modern human, a presumption confirmed by other bones. The Australopithecine footprints were undeniably "human" and they provided us with the earliest date we have for the emergence of competent bipedality as a preferred mode of locomotion. What these impressions also finally resolved was the argument about which came first: big brains or bipedal locomotion. Studied alongside research on the skulls and limbs of Australopithecines, they confirmed that it was, indisputably, walking on two legs, and the freedom it gave us to use our upper limbs to explore, that first characterized us as human. Perhaps only then did we start to work on our big brains. Standing upright was the pivotal action that changed the future of our species, other species and our planet. As Wood Jones had insisted, we owe it all to the humble foot.

Other countries might not be able to match the richness of the palaeontological treasures of Africa, but one of the oldest sets of hominid footprints found so far outside that continent was discovered in the UK. The Happisburgh footprints, made by a group of adults and children, were revealed in 2013 in Norfolk, in the muds of an ancient estuary, and dated to between 850,000 and 950,000 years ago. They were stumbled upon by a team of scientists who were working on another project after the protective layer of sand that had been concealing them was washed away in the huge St Jude's storm of that autumn. The sediment lay below the high-tide mark and the scientists knew they were in a race against time and tide to record them before the sea eroded them permanently. Their swift thinking won them a Rescue Dig of the Year Award after their pictures were exhibited later at the Natural History Museum. Within two weeks of re-emerging, the footprints had gone.

Footprints, and what we can tell from them, fascinate scientists across many different specialisms. While clinicians will look at them to see if there is an abnormality they may be able to fix, forensic podiatrists study them to compile evidence for the court. Perhaps a footprint has been left in blood at a crime scene, or in soil outside a window, and it may be possible to match this to a suspect. Obviously, this investigative approach has greater value in situations where people habitually walk around in bare feet. In cooler climates, and outside the home, it is much more likely that we will find shoe prints.

But these can be useful, too. Shoes can be matched to the person they belong to, especially when they have been worn without socks or tights. If you look inside one of your shoes, you will see some kind of replication of your footprint. A podiatrist could compare this print, or at least a version of it, with your foot to determine the likelihood that the shoe and the print are yours.

Footprints can give us quite a lot of information about the person, or people, who left them. For example, we can estimate the length of their stride and therefore their height, just as it was possible to do with the Australopithecine impressions. We can work out what shoe size they take. We can tell how many people were present at a scene and whether they were standing, walking or running.

If the print of a bare foot is sufficiently clear, we may be able to lift toe prints in the same way as we do fingerprints. These were of some help in identifying the bodies of children after the Asian tsunami of 2004. Toe prints could be compared with bare footprints found around the family home where, say, the child might have climbed on furniture. More recently, Japan has been considering setting up a footprint register alongside their fingerprint database. This may sound a strange idea, but there is a logic to it. Because feet are frequently protected by shoes, they tend to survive better in mass fatality situations than other parts of the body. For this reason, the records of some military air personnel may include bare footprints as a potential additional means of identification in the event of a plane crash.

In recent years, probably the most infamous case involving footprint evidence was the murder of Meredith Kercher in Perugia, Italy in 2007. The body of twenty-one-year-old Meredith, a British exchange student, was found on the floor of her bedroom in the flat she shared with three fellow students. One of her flatmates, Amanda Knox, and her boyfriend, Raffaele Sollecito, were charged with Meredith's murder and a third person, Rudy Guede, a regular visitor to a neighbouring flat, was also later arrested in connection with the crime.

With three defendants, it was always going to be difficult to separate truth from speculation, and at the centre of much of the confusion was some less than reliable forensic evidence, including a partial footprint in blood on a bath mat at the scene. The blood was confirmed by DNA analysis to be Meredith's; the owner of the footprint was not so easy to determine.

The prosecution alleged that the print was a "near perfect" match with Sollecito's right foot, but not for Knox or Guede. However, expert witnesses called by the defence pointed out fundamental errors in the testimony of the prosecution's expert and offered evidence that the print was more likely to be Guede's. The prosecution's witness was a physicist, not an anatomist, and it is always troubling when scientific evidence relating to anatomical features is being interpreted by a professional whose expertise is in another discipline.

The custody footprint taken for comparison with the print from the scene was static and had been recorded in ink on paper—two very different materials from the blood and thick fabric involved in the formation of the original print. No attempt had been made to replicate the effects of the much greater absorbency of the bath mat or the consistency of the blood.

Guede opted for a fast-track trial and was found guilty of the sexual assault and murder of Meredith. He was sentenced to thirty years in prison, later reduced to sixteen. Knox and Sollecito were convicted of murder and both served almost four years in jail before being acquitted on appeal. The appeals were then quashed and both were found guilty a second time, only for these convictions to be annulled once more, by the Court of Cassation, the highest court in the land, on the grounds of reasonable doubt. This decision definitively ended the case and Knox and Sollecito walked free.

The pattern of footprints can tell us whether the person who left them was standing still or moving. We can all recognize people by the way they walk, although we are usually processing other clues simultaneously. Despite my poor eyesight, I can pick out my husband from a distance by how he stands and walks, but I am also going by his size, shape and the clothes he is wearing and, more often than not, because he is roughly where I expect him to be, even if the figure I am looking at is all a bit of a blur.

Gait analysis, the study of the manner in which we move, is quite distinct from this kind of everyday recognition. Experts in this forensic technique claim to be able to match the pattern of movement of an offender—often with nothing more to go on than some very poor-quality CCTV images taken from an odd angle—with that of a suspect in front of them in a police station custody suite. Both offender and suspect, if indeed they are two separate people, will usually be unknown to the expert and probably wearing different clothes, so the rationale is that comparison is being made on gait pattern alone. But the fact that these are very different environments could have a bearing on the way a suspect moves. In the first, they are unaware they are being watched; in the second, they know that their walk is being scrutinized.

It is said that our walk is unique to each of us, but there is no solid evidence to support this theory. Of course, if someone has a particularly unusual gait, such analysis is likely to be more reliable, but the way we move is not necessarily always going to be the same. We walk differently if we are in high heels from how we walk in flats; if we are wearing comfortable or uncomfortable shoes; if we are carrying a heavy bag on one shoulder, or a bag in each hand, or if we are walking uphill on cobbles rather than downhill on an even pavement. We do not yet have enough valid information on how our locomotion may be affected by these and other conditions.

Gait analysis has been presented in court to convict defendants but as the methodology is relatively new, care must be taken over the safety of the evidence. The Rt Hon Sir Brian Leveson summed up the need for caution when he described forensic gait analysis as "a much younger and less scientifically robust area." A "judicial primer" has now been given to all judges in the UK to clarify for them where the science for gait analysis is reasonably tried and tested and where there is much research still to be done.

In 2013, expert evidence presented by a forensic podiatrist was used by the defence as the basis of an appeal against a murder conviction. Following an altercation outside a McDonald's restaurant in Wythenshawe in 2006, a twenty-five-year-old man had been shot dead. The case against the alleged gunman had collapsed and he had been acquitted. But if you are party to a murder, you do not have to be the one who pulled the trigger to be charged with the crime, and Elroy Otway, the man accused of being the driver of the getaway car in which the gunman had been a passenger, was tried in 2009 on the basis of "joint enterprise," found guilty and sentenced to a minimum of twenty-seven years in prison.

The car had been identified and CCTV footage from a service station showed a man filling it with petrol shortly before the murder. The forensic podiatrist called to give expert evidence had compared Mr Otway's gait in the custody suite with that of the individual on the CCTV recording.

At appeal, the defence counsel argued that gait analysis was not sufficiently advanced as a methodology to be permitted as evidence and they did not accept that the podiatrist was a competent forensic expert. The evidence was, they claimed, circumstantial. However, the three judges hearing the appeal in London considered the evidence as a whole and overruled the grounds for the appeal. The trial judge had, they said, been entitled to rule the evidence as admissible, and to allow the court to hear the podiatrist's opinion, leaving the validity of the forensic gait analysis open to debate. They did add, though, that they did not endorse the use of podiatric evidence in general. It is important that scientists and the judiciary work together to ensure that what gets into court to be heard by the jury is founded in verifiable science, which meets the required standards of repeatability, reliability and accuracy, and that it is probative.

Running produces a different characteristic human gait and footprint from walking or standing. Standing requires two feet to be in contact with the ground. When we walk, one foot at a time leaves the ground. At the peak of a fast run, there is a phase when there is no contact with the ground at all and the runner is technically airborne. The distinction between walking and running is central to the rules of speedwalking, in which running is prohibited. Hence the rather odd gait synonymous with the sport, which enables the competitors to walk extremely fast while always keeping one foot on the ground.

Human walking has a double-pendulum action called the gait

cycle, which involves both a standing and a swing phase in each leg at different times. The stance phase occupies about 60 per cent of the cycle and the swing phase the remaining 40 per cent. Gait involves a combination of movements in a chain across both phases. At any one point in time, a limb is in one of the following positions: heel strike, foot flat, mid-stance, heel off, toe off and swing. Try it. Walk in slow motion and note where each limb is during the different phases of the cycle.

The stance phase begins with heel strike and the swing phase with the toe off. The whole of the foot becomes engaged in the walking motion, from the heel at the back to the big toe at the front. This is why, in a walking footprint, the deepest impressions are made by the heel when it strikes and the big toe when it pushes off. In a purely standing footprint there is no "dig-in" associated with the heel or the big toe.

Although our feet have only two main functions, keeping us standing and moving us around, we can train them to become incredibly dextrous when necessary. Indeed, Luther Holden, a nineteenth-century anatomist surgeon from Birmingham, described the foot as "*pes altera manus*," loosely translated as "the other hand." The bones are homologues of those in the hands, with the seven tarsals in each foot equating to the eight carpals in our hands and the five metatarsals to the five metacarpals. And the phalanges, of which we have fourteen in each hand and each foot, have the same name and position in the toes as they do in the fingers: distal, middle and proximal. Other than our big toe, the hallux, and our little toe, digiti minimi, we simply number our toes from one (big) to five (small) without bothering to give them names.

Admittedly, the foot does not have the full agility of the hand. As the foot has no comparable equivalent of the opponens pollicis muscle, the pads of the smaller toes cannot be brought together in a pincer action with the big toe as the fingers and thumb can, and the big toe therefore occupies a very different position from the thumb. But that aside, all the other equivalent muscles and bones are present to provide us with the technical capability to use the foot as a substitute for the hand should the need arise.

History is peppered with the names of those who have shown that the loss of their hands as a result of illness, accident or congenital disability need not be a barrier to creative art. The celebrated four-teenth-century German artist and calligrapher Thomas Schweicker lost both his arms in a duel over the right to court a lady. The skills he cultivated eventually attracted the attention of the holy Roman emperor Maximilian II, who brought him to the royal court. One of his works, a self-portrait which was reproduced on his tombstone when he died in 1602, shows Schweicker writing with a brush held between the first and second toes of his right foot while using his left foot as a guide.

In 1957, the British Mouth and Foot Painting Artists (MFPA) self-help partnership was set up by a small group of artists from Britain and eight other European countries who painted without the use of their hands. It is still going strong today. Christy Brown, famous for his book *My Left Foot*, later turned into an Oscar-winning film, was an early member of the group. Probably the best-known mouth-and-foot painter in the UK now is Tom Yendell, born a "thalidomide baby" without arms. He says simply: "I learned to adapt"—a beautifully suc-cinct summary of the extraordinary capacity of the human body to reinvent, almost to reset. That as a species we can find the capability in our bodies to adapt to such an extent is nothing short of miraculous.

But the big toe will, of course, always play second fiddle to the thumb in the majority of us fortunate to possess both. As the loss of a thumb will have a far greater impact on our daily lives than the loss of a big toe, the transplantation of the hallux to replace an amputated pollex has become a recognized surgical procedure.

The first foot-to-hand transfer was performed in the UK in 1968. The patient was a woodworker whose thumb and first two fingers had been sliced off in an accident with a circular saw. The substitution of his missing thumb with his big toe successfully restored some of the dexterity in his hand. Surgeons will usually connect at least two nerves, along with the corresponding vessels, muscles, tendons and skin, and the transplanted digit, sometimes referred to as a "thoe," has proved to be very effective compared with an artificial prosthetic which, however good, lacks the subtlety of movement and sensitivity of real skin and bone.

These patients learn to do without their big toe, but for some, it would seem, it is a loss that cannot be tolerated. The earliest toe prosthetic we know of was crafted from three pieces of hinged wood and leather, with a carved and sunken nail. Nicknamed the "Cairo Toe," it dates back to between 1069 and 664 BC. It was found in a necropolis west of Luxor, with the remains of the Egyptian mummy Tabaketenmut. Jointed in three places and constructed to fit its owner perfectly, it was likely to have been reworked several times as she aged.

Tabaketenmut, the daughter of a priest, was probably between fifty and sixty years old when she died. Evidently she had suffered an amputation of her right big toe at some earlier stage in her life, perhaps, it has been suggested, due to gangrene or diabetes. Her foot had fully healed but for some reason she wanted to disguise the deformity. Was it simply vanity? It has been postulated that this was to aid her balance, but the absence of a big toe causes no significant problems in this department. Even a Pobble amputation (disarticulation across all the metatarsophalangeal joints, which removes all the toes) has limited impact on balance, walking or standing. It is only swift movements, such as running, that prove difficult. And, as a priest's daughter, it is unlikely that Tabaketenmut was a sprinter.

Of course, all manner of items were buried with mummies in Egyptian tombs for their use in the next world, so it is possible that the prosthetic was created solely for burial or ritual purposes to ensure that Tabaketenmut did not go to the afterlife incomplete. However, evidence of wear and tear, together with the probability that it had been altered more than once, suggests this was not simply a funerary adornment. Perhaps she just wore it so that her sandals fitted properly.

Another, more recent, prosthetic right toe from Egypt, named the Greville Chester toe after the collector who acquired it for the British Museum in 1881, dates back to before 600 BC. This was made of cartonnage—multiple layers of linen or papyrus impregnated with animal glue—a composite more commonly used for the construction of mummy cases. As the Greville Chester toe does not bend it is likely to have been purely cosmetic. It has a cavity where the nail should sit, which was perhaps inlaid with a different material, either to make it look like a more authentic nail or maybe to show off an early example of nail bling.

◊

The average length of a baby's foot at birth is about 3 ins (7.6 cm). It will grow rapidly in the first five years of life as it must mature swiftly to take up its functional role. By the end of the first year it will already be nearly half the length of an adult foot and by the end of the fifth, it will measure around 6 ins (15.2 cm).

Most children will have adopted a shaky bipedality by around ten to sixteen months, but a fully mature gait will not be mastered until approximately six years of age. The foot will keep growing until about thirteen in girls and fifteen in boys. Interestingly, although the upper limb and hand appear in the embryo before the lower limb and foot, it is the foot that reaches adult size ahead of the hand. This is because the requirement to develop a stable foot takes priority.

Parents tend to buy that all-important first pair of shoes within six to eight weeks of their child's first independent steps. We know, however, that to aid healthy development, the more the foot is left bare and unshod, the better. About 5 per cent of the population visit a podiatrist or a chiropodist every year with some foot-related complaint, at the root of most of which are poorly fitting shoes. Women are the worst offenders. They will often buy shoes for their aesthetic appeal, or to complete an outfit, rather than for comfort or health. Platforms, wedges, stilettos, winkle-picker toes, ballet flats, flip-flops and many other fashionable styles are all basically torture chambers for the feet.

The long-term effects of ill-fitting or inappropriate shoes, and of activities for which the foot was never designed, can be sobering. My daughter once asked a chiropodist what were the worst feet he had ever seen and he replied, without hesitation, that they were the feet of an elderly ballerina, which looked like two plates of rice pudding. His words, not mine.

There is a correlation between height and shoe size, with taller people tending to have larger feet. So it is no surprise that the biggest living adult feet in the world belong to an avid basketball player, Jeison Hernandez from Venezuela, who is 7 ft 3 ins (220 cm) tall. In 2018, when he was twenty-two years old, his left foot measured 40.47 cm in length and his right 40.55 cm (nearly 16 ins). He takes a US shoe size 26 (UK size 24). The smallest adult feet may belong to Jyoti Amge, a young Indian woman who is just 24.7 ins (62.8 cm) tall. Her feet are 3.5 ins (9 cm) long, around the same size as those of a one-year-old.

The notion that small is beautiful as far as feet are concerned was, of course, taken to its most extreme by the Chinese custom of binding women's feet, which persisted from the tenth century right up to the early years of the twentieth. Bound feet were at one time considered a status symbol as well as an ideal of beauty. Known as "lotus feet" and encased within wrappings and tiny "lotus shoes," they were viewed by some as the most intimate and erotic part of a woman's body.

Upper-class women, to heighten their allure, would soak their feet, cut the nails and then bind the toes tightly into the sole of the foot. With the toes curled underneath, the foot would be pressed down with great force until the toes and the arches were broken. Ultimately, the bones would heal in this abnormal position.

The effect was to bring the ball of the foot and the heel together so that the middle part of the foot was raised. Feet were often unbound and rebound daily to remove necrotic tissue and bones might have to be rebroken if they were not healing in an aesthetically pleasing way. The tight binding resulted in poor circulation, infection and constant pain. Sometimes toenails would be removed entirely. And if your toes dropped off through gangrene, this was seen as a bonus. The perfect lotus feet would be no longer than four inches (10 cm), the foot size of an average toddler.

It goes without saying that any form of locomotion was challenging for these women. So, too, was standing. We find it difficult enough to stand upright on our normal-sized foot pads for long periods of time. That is because it takes incredible co-ordination across our mus-culoskeletal system to stop us falling over. Workplace advice acknowl-edges that standing demands around 20 per cent more energy than sitting and recommends that we should not stand still for more than eight minutes at a stretch.

If you try to stand on one leg for any length of time, it becomes very clear how precarious the balancing act of standing actually is. And if we add some inebriant into the mix, we lose our balance easily because this affects our ability to control the intricate co-ordination needed to maintain our equilibrium. When we are balanced, our line of gravity passes from in front of our spine to behind our hips, then just in front of our knees and ankles and down to a base of support between the feet that is just a few square centimetres in size.

Sheathing our feet in socks and shoes for protection and warmth is a very human characteristic. For the anthropologist, these coverings can be most useful. Traditional natural materials such as wool, leather and skins have now been supplemented by the synthetic equivalents of the modern age. They all help to keep the component parts of the foot together, even when the rest of the body starts to disintegrate, as well as sometimes conserving them better. A shoe also makes it diffi-cult for a predator to remove a foot. And if a body ends up in water, it can act as a flotation device.

A strange series of events occurred between 2007 and 2012 in the Georgia Strait, which separates Canada from the US. During these five or six years, twenty separate feet, inside their shoes, were washed ashore. A shod foot can float for a thousand miles and the cold tem-perature of the water will turn the fat in the foot into adipocere, the wax-like substance formed by anaerobic hydrolysis, lending more buoyancy to the foot and helping to preserve its soft tissue. Some of these feet were matched to missing individuals, but of course the wild stories and imaginative myths that grew up around the phenomenon took the internet by storm.

At the height of the frenzy, one bunch of students decided to stuff a decomposing animal foot into a sock, put it into a trainer packed with seaweed and leave it on the shore to be found. But as we all know, the human foot is not like any other, and it did not take the anthropol-ogists long to rumble the hoax.

People are lost at sea all the time—in boating accidents, plane crashes or other mass fatality events—or they may be deliberately buried at sea. As the body decomposes in the water it may naturally separate into its constituent parts. When a foot has its own buoyant vehicle to keep it afloat, it is not surprising that it will move with the tides and eventually be washed up on a shoreline.

The UK has about 7,723 miles of main coastline (if you don't count the islands), about half of it owned by the Crown estate. So finding iso-lated feet on UK beaches, and along riverbanks, the shores of lakes and lochs or in canals, is not unusual.

One right foot, wearing a training shoe, was discovered in a river on the east coast of England, shortly followed, further upstream in the same river, by a left foot. Only this one was wearing a brown boot. The booted foot was traced to a man who had gone missing that same year, while the right foot in the trainer belonged to an entirely different gen-tleman who had disappeared two years before. His left foot was even-tually identified as well after it turned up on a beach in Terschelling, one of the West Frisian islands off the north coast of Holland, having made its way across the North Sea.

From single feet like these, forensic anthropologists can deter-mine the sex (from the size, the presence of hair, and so on), age, height and shoe size of the individuals from which they have become separated and sometimes this information can be sufficient to create a broad identity profile that assists with narrowing down the possibili-ties. But on its own, it won't be enough to pinpoint a positive, named identity.

Interestingly, finding a single, isolated foot is generally considered insufficient grounds for opening a coroner's inquest as it cannot be taken as indicative of a death. Although the severing of a foot is quite likely to be a life-ending event, it is, of course, possible for someone to survive such an amputation.

It is rare for foot bones to be the only evidence in an investigation but I do remember one, long ago, when I was still working in London.

The call came from a police officer in Cambridge, who had a case out in the fens relating to the remains of a Polish Second World War pilot and his aircraft. The pilot had been returning to the UK after a sortie across the North Sea in about 1944, if memory serves, when his Spitfire took a direct hit to its engine, which eventually failed as he passed over the east coast. He had no time to eject. His plane plunged nose-first into the flat wetlands, its wings ripping off, reducing it to a metal tube like a cigar case embedded in the boggy ground.

The sites of these incidents are well documented and the authorities know where most of them are. Occasionally, bits of plane, or indeed of pilot, have come to the surface as farmland is ploughed year in, year out. This request came around the end of February, early March, as I recall, a time of year when farmers are starting to think about preparing their fields for the sowing of new crops.

There is big business in wreck salvage, and a lot of money to be made from Second World War aircraft artefacts. The police were aware that salvage-hunters had been searching this area for a couple of years and the military knew there had been various small finds. But the prospect of unearthing an intact Spitfire fuselage, preserved by the fenland soil, would have made these fields a prime target for those perhaps more interested in their monetary value than in their historical importance or the sanctity of a war grave.

It was one of these salvage teams that had contacted the police. They said that, while walking in the field where they believed this Spitfire had crashed, they had come across a bone which they thought might be human. They had left it in the field where they had found it, flagged by a marker. They realized it might be important and felt that it should not be moved.

The police were suspicious. They'd had dealings with this group before, and knew they had previously found an aviator's boot with some bones in it. They would not be surprised, they told me, if this bone (if that was what it was) had come from the same source and had little to do with the fuselage they were looking for. The salvage hunters were perfectly well aware that if human remains were discovered, there would have to be an archaeological and anthropological search and assessment of the site, which might well lead to a full-scale excavation, and an opportunity for them to claim the salvage.

It was a bitterly cold morning when I arrived in a police car at the side of the newly ploughed field to walk the area with the search officers. An orange camping flag had been left in the field for us, marking the spot where this bone had allegedly been found. We headed there first, so that before we did anything else I could determine whether it was actually human.

There was no doubt about it. It was a human left fifth metatarsal, the bone at the base of the little toe. What made me uneasy was that it was sitting on top of the soil, not partly buried in it; not even adhering to the wet earth beneath it. It was clean, with no sign of dirt, dust or mud. It looked as if someone had deliberately placed it there. We photographed the bone, lifted it and bagged it as evidence. Then, starting at the edge of the field, we worked our way inwards, walking every drill of the ploughed area. We found no trace of a boot or shoe, no sign of any wreckage. Not even any animal bones.

What we did find, within a radius of less than eight feet of where the fifth metatarsal had apparently been discovered, were another four small human bones, all from a left foot. When compared with the metatarsal, they all looked, from their size, colour and appearance, to have come from the same individual. But every single one of them was perched right on top of a plough ridge. Like the metatarsal, none were buried in the soil, and there was no soil sticking to them. In my view, they had simply been laid on the ridge to get our attention.

The salvage crew, of course, professed to know nothing about this, or about where the bones might have come from. But if their intention had been to coerce us into an excavation, they failed. The police and the military authorities accepted my opinion that the bones had probably been left there on purpose for us to find and my advice was that these were not justifiable grounds for an excavation.

I did suggest that they attempt to retrieve some DNA from the bones, in case it was possible to establish a match with any of the pilot's relatives. But at that time DNA extraction was not as sophisticated as it is today, the bones were all very heavily weathered and the lab was unable to obtain sufficient genetic material. I think the remnants were buried as unidentified. And what was left of the Spitfire and the pilot remained, for the time being at least, in their quiet fenland grave.

It is not only the bones of the feet but my personal bête noir—those deplorable toenails—that can tell us something of the life lived. The toenails grow at the rate of about a millimetre a month (significantly more slowly than the fingernails) and it takes them approximately twelve to eighteen months to fully regrow. The average toenail, then, represents a record of perhaps the last couple of years of a person's life, with the nail bed holding the most recent information and the tip of the nail the oldest. If you know what you are looking for, there is a tremendous amount that science can tell from that nail in terms of where a person lived and what they ate and drank.

The work of Professor Wolfram Meier-Augenstein and his team on the role of stable isotope analysis in human identification, and its forensic application, has been groundbreaking. It was his expertise in a case of the death of one young boy that helped to convict the child's father.

Paramedics had been called to an address where a pitifully thin little boy was reported to be unresponsive. Noticing some bloodstains on the banister of the staircase and a spherical indentation in the plasterwork, they informed the police. The child died in hospital and the postmortem report identified multiple injuries to his brain and internal organs. There was also a list of old injuries. The father denied ever having hit his son or being involved in any way in his death. He was, he claimed, a loving parent.

Tissue samples were sent for analysis, including some pieces of bone, a big toenail, a thumbnail and some muscle. The big toenail provided a chronological profile of the child's nutrition over the previous year.

Professor Meier-Augenstein was able to identify three periods of dietary life history and two changes in dietary lifestyle. The period most distant from death, four to twelve months before, was nutritionally stable, and the little boy was consuming a normal omnivorous diet. Four months before he died, there was a shift towards a different diet, based on plants in the C3 category (such as wheat, rye, oats and rice) rather than C4 plants (maize, sugarcane and millet). One explanation for a change like this would be a move from a warm and arid climate to a more temperate one. In the last two months of his life, there had been a pronounced departure from an omnivorous diet to one containing virtually no animal protein.

The police had learned through their inquiries that until about four months before his death, the boy had been living in Pakistan with his mother, without any physical contact with his father. The child had then come to the UK, where he was in the parental custody of his father, although initially he had lived for the most part with his grandparents. But for the final two months of his life, his father had been his sole carer. The grandparents had been told that his bruises and injuries had been accidental and had no suspicion of what was happening to their grandson. Eventually, the boy's father admitted manslaughter and he was given a nineteen-year prison sentence.

Our bodies do not lie, but sometimes it takes an expert to coax the truth from them. Even from our toenails.

Feet may hold other information about our lifestyle. For example, tiny pinpricks can alert the forensic anthropologist to intravenous drug use. The foot is not, of course, normally the preferred site for injecting drugs—that is usually the upper limb—but within about four years of intensive drug abuse, the veins may start to collapse and become inaccessible, and the habitual user will often move on to the leg and foot. Because there is so little soft tissue covering the back of the foot, the veins here are often very visible and therefore easy to locate. Evidence of the habit can also be conveniently hidden away under socks and shoes.

Many IV drug-users choose the vein between the first and second toes, but injecting here carries an increased risk of complications such as slow healing, abscess, infection, vein collapse, thrombosis and leg ulceration. The veins in the foot are thin and prone to bursting under injection pressure. Heroin addicts often use tattoos to hide injection sites and track marks and so when we see tattooing on the feet, especially between the big and second toe, it signals to us that the foot will merit closer inspection.

Because of the volume of nerve endings in the feet (just think how ticklish they are), pain we experience here can be acute and debilitating. And that makes them a target for anyone wishing to inflict it. Falanga, or foot whipping, an ancient practice now recognized as a form of torture by the European Court of Human Rights, has been documented in various parts of the world, mainly among Middle and Far Eastern nations. It can result in walking impairments and sometimes in fractures to the bones of the feet.

In January 2014 I was contacted out of the blue by an international legal firm based in London. They were putting together an independent investigatory team that included international litigators and forensic practitioners. Would I be prepared to fly out to Qatar and examine some images? I would be away for no more than a week and all my costs and expenses would be covered. An air ticket would be left for me to pick up at Heathrow airport on the appointed date. I was not told what these images contained or who else would be in the team.

Being a cautious sort, I checked out the firm. They seemed perfectly legit and the gentleman who had contacted me was clearly a highly regarded international lawyer. I also checked with my Foreign & Commonwealth Office contacts, who assured me that they saw nothing that concerned them in their approach.

But this was as much as I knew. Other than that, I was on my own, and could potentially be putting myself out on a limb. Hearing nothing further from the law firm, I decided to leave it to fate. As I was in London anyway, I would go to Heathrow, and if there were no tickets at the desk, so be it. If the tickets were there . . . well, perhaps it was time for a new adventure.

My tickets were waiting for me—and they were first class. I wasn't sure whether to be reassured or worried. But I must admit that it is lovely to be able to turn left as you get on to a plane, and while I have occasionally been granted the luxury of travelling business class, I was about to discover that first class is on a different level altogether.

Unfortunately, I was still recovering from a particularly debilitating, month-long bout of labyrinthitis, an inner-ear infection, and I was anxious about flying. Moreover, I was not touching alcohol, in case it interfered with my medication, and being very careful about what I ate, for fear that the overwhelming motion sickness associated with the condition might return. So I sat there in my spacious first-class seat, for what was the first and, I suspect, will be the only time in my life (I am essentially a budget girl), rejecting all offers of champagne, fine wine, scallops, steak and chocolate and accepting only bread and water. But it was a wonderful experience all the same. The beds were very comfortable, the attendants most attentive and the free gifts all by Dior. Still, I couldn't escape the feeling that I was an imposter who might be caught out at any moment.

When we landed at Hamad International airport in Doha, all first-class passengers were asked to remain on board. It appeared that there was a limousine assigned to each of us to take us to the terminal. I could get used to this kind of travel, I thought to myself. Then my name was called and I was requested to stay until last. I knew it—found out at the last minute, and now I would have to face the solitary walk of shame to the terminal.

But no, it turned out that I had a special limousine with a senior politician to escort me through the airport and to my hotel. He took my passport, spoke to immigration, had it stamped and collected my luggage for me while I just relaxed in the car. Yet there was a prickle of disquiet at the back of my neck. When a government goes to these extremes, there may be a price to be paid, so you must remain vigilant. There is, as they say, no such thing as a free lunch, and this standard of treatment was highly unusual, to say the least. I was driven to an incredible hotel and shown to my own private suite. An entire floor had been given over to our team so that we could work there in seclusion. It was very clear now that whatever we were here to do carried the Qatari government's full backing. There were six of us. Three of our number were among the world's most eminent international criminal lawyers. Also in attendance was the lawyer who had approached me to begin with, who turned out to be the most charming of gentlemen. I was also very reassured to encounter a pathologist from the UK who was a friend of mine. And then there was me. We were briefed by a government official.

The Arab Spring protests of 2011 had led to significant unrest in Syria against the government of President Bashar al-Assad. A number of the demonstrations there had been violently suppressed and many men had gone missing or been detained. *The Financial Times* had reported that Qatar had funded the Syrian rebellion by "as much as $3 billion" over the first two years of the ensuing civil war, and that the country was offering refugee packages of about $50,000 a year to defectors and their families. One such defector was a man only ever known to us by his code name, "Caesar."

Caesar told war crimes investigators that before the Arab Spring he had been a forensic investigator. Once the Syrian uprising began, he had been employed as a photographer in the Syrian military police. His job was "taking pictures of dead detainees" at two military hospitals in Damascus, documenting the corpses of those who had died in Syrian military prisons. He did not claim ever to have witnessed executions or torture, but he did describe a highly organized system for recording deaths.

He alleged that every body brought to the hospital was marked with two numbers: one purported to relate to their hospital admission and the other was their detention camp number. He had to photograph each face alongside its hospital number, an image that could be sent to the man's family, with a message regretfully informing them that their son/husband/father had died in hospital of natural causes and that the photo was the proof required for a death certificate to be issued.

He never saw evidence of torture on these faces. All the evidence of torture was below the level of the jaw. These injuries also had to be photographed—this time with the detention number, as proof that the orders given for the prisoner to be tortured had been carried out.

Caesar became increasingly affected by the scale of the operation unfolding in front of him and, seeing a way out in the tantalizing refugee package being offered by Qatar, he began to talk to a rebel group. He was persuaded, at great personal risk, to copy the photographic evidence. Every day he brought out a USB stick, concealed in the toe of his sock, and passed it to the group.

They managed to smuggle these USB sticks out of the country and, in August 2013, Caesar "died" in a tragic road accident. In reality, while his grieving family attended his funeral, he, too, was being secretly transported out of Syria. By the time we met Caesar in Doha in January 2014, his family had also been successfully removed to safety and reunited with him.

He had copied over 55,000 images of what were claimed to be more than 11,000 bodies, all showing evidence of starvation, brutal beatings, strangulation and other forms of torture and killing. The task of the lawyers on our team was to interview Caesar to determine whether he was a credible witness, while the forensic pathologist and I were required to examine as many of the 55,000 images as we could to establish whether they were fabricated, whether any of them were duplicates or whether they were the real thing.

The inquiry team approached the job of evidence evaluation with caution. We were all alert to the need to guard against being used as a vehicle for the advancement of any particular political point of view. It was vital that the team came to its own conclusions, that we were operating without interference from other parties and that our findings were unbiased and justifiable.

The pathologist and I left the hotel in a car which took a very circuitous route to an apartment block in downtown Doha. We would visit that apartment on three separate occasions, but we never went the same way twice. We were aware that we were being followed by Qatari special forces to ensure that the secrecy of Caesar's location was maintained.

We sat in the car until given the all-clear to enter the building. At the door of the flat we were met by security, who checked that we were "clean" before we were taken into a very sparsely furnished sitting room and left there alone for quite some time. Finally, Caesar entered the room and we were introduced. He was a quiet and likeable man. A laptop was brought in. We were allowed to open it and the folders it held, which contained thousands of images, all of deceased males. We spent the first hour simply flicking through these, to become accustomed to what we were seeing and also to look for signs of any evidence of duplication or staging.

Caesar was, not surprisingly, very wary of us at first, but as we spent time with him, and he was reassured that there was no suspicious agenda to our questions, he relaxed. We asked if he had taken all the images himself. He said that he had not. We asked if he had witnessed any of the killings and he confirmed that he had not. It appeared from his responses that he was not trying to exaggerate any of his claims.

In one picture there was a clear image of the photographer's thumb. Although I knew it was not Caesar's thumb, because I had done an assessment of his hand anatomy while sitting next to him, I asked him if it was. He said that it was not. No matter how many times we posed our questions from different angles, his answers were always consistent and unambiguous, and if he didn't know the answer, he was content just to say so.

We talked about the report we would be writing, although at that time I was not aware of its ultimate purpose. The men surrounding Caesar were insistent that we could not use any of the photographs in the report but I urged them to rethink, reasoning that the visual effect of what we were seeing was particularly important. I showed them how we could anonymize any images we were permitted to reproduce by blacking out faces and numbers in such a way that these changes could not be reversed. After much discussion and debate, they eventually agreed to us using a few of the images—no more than ten—as long as their subjects were completely unidentifiable. This was a significant victory, as it transformed the impact and the value of the report.

It was clear that we were not going to have sufficient time to examine all the photographs in the files, so what we decided to do was to "dip-sample" into every folder to gather a broad representation of the nature of the injuries that we were seeing. In total we worked through nearly 5,500 of the photographs and sorted them into categories.

Like Caesar, we saw no evidence of torture in the bodies above the level of the jaw. But sixteen per cent of our sample showed transverse ligature marks around the neck. These were inconsistent with hanging, in which the imprint left by suspension generally turns upwards towards the back. It was our opinion that these were strangulation marks, and consistent with torture. In one picture the ligature used, a fan belt from a car, was still around the individual's neck. Some ligature marks were present on wrists and ankles and, in one case, a plastic cable tie was visible.

Five per cent contained what are known as tramline bruises: where a person is struck with a long, cylindrical object such as an iron bar, or even a plastic rod, which bursts the surface of the skin, causing linear open bruising. These very distinct sets of parallel marks were seen not only across the torsos of some individuals, but in other cases on the limbs as well. One body in particular showed so many such bruises—over fifty—up and down the torso, that the victim must have been bound at the time, otherwise he would have been trying to curl up to protect himself.

There was a high level of emaciation in over 60 per cent of our sample. Many of the bodies were so thin it was as if we were looking at photographs from the concentration camps of the Second World War. Bones protruded through skin, every rib was visible and faces were hollow and sunken.

The last category of specific trauma we analysed related to the lower limbs, and specifically to the shins and feet. Over 55 per cent depicted extensive ulceration to this area of the body. The precise cause was unclear and Caesar was unable to enlighten us as he only ever saw the aftermath of the torture. Potential explanations include pressure effects (pressure sores), vascular insufficiency, inflicted injury, such as application of hot or cold objects, and tissue breakdown resulting from poor nutrition. But with the majority of the ulcerating lesions in this sample occurring in young men, a natural cause for all of them was highly unlikely.

The most probable explanation was venous insufficiency as a result of a cripplingly painful torture involving a ligature bound tightly around the knees that severely restricts the movement of blood in the lower limbs. The build-up of pressure sees vessels rupture and skin ulcerate. More than half of the men we sampled had this in both their legs and feet, and it was evidence of a regime and pattern of torture.

The photographs were not always taken from the best of angles, from the perspective of a forensic scientist trying to accurately diagnose some of these torture injuries, and the punishment of falanga could not be excluded. Falanga beatings, which have been reported in Syria, tend to focus on the soft, middle-arch region of the foot, rather than the heel or the ball, and require either co-operation between torturers or for the victim to be immobilized—forms of restraint that some images suggested were being used while inflicting other injuries.

We found Caesar to be a credible and compelling witness and the images to be genuine. Our report, the Da Silva report, was completed in Qatar. By the following week its publication would be covered by newspapers and broadcasters all over the world, many of them, including the *Guardian* in the UK, reproducing it in full. It was timed to coincide with the Geneva II Conference on Syria, a United Nations–backed international peace conference aimed at ending the Syrian civil war. The launch of the report, on the eve of the peace talks, put Assad's regime on the back foot and sparked international outrage and condemnation of the industrial scale of the slaughter. But as yet, this has led to no obvious resolution.

It feels appropriate to leave the last word to Caesar. He told US President Obama: "I have risked my life, and the life of my immediate family, and even exposed my relatives to extreme danger, in order to stop the systematic torture that is practised by the regime against prisoners."

# Tailpiece

*"You don't know that you'll ever have to talk about the skeleton in your closet"*

Mark McGwire
Baseball Player

The American epidemiologist Nancy Krieger summed up better than I ever could the relationship between our bodies, ourselves and our world when she wrote of how the stories that our bodies narrate cannot be divorced from the conditions of our existence. They often match our own and other people's stated accounts—but not always. Our bodies tell the stories that others cannot, or will not, tell because they are unable, forbidden or have chosen not to do so. Since *All That Remains* was first published in 2018, a remarkable number of people have written to me to tell me about their own bodies: about what has gone wrong with them over the years, the weird and wonderful anatomical variations to which they attest and what their remains might look like after they have breathed their last. Together these stories form an incredibly rich tapestry of the sheer range of human anatomy that exists in our species and are testament to just how unreserved we can be about sharing them.

*Written in Bone* focuses on the body section by section because that is how a forensic anthropologist works. We have no way of knowing what part or parts of a body might be presented to us for identification or in what state of preservation or fragmentation. As all of the cases in this book illustrate, our job is to squeeze every single piece of information out of whatever parts we do have in our pursuit of the answers to questions about identity, life and death.

The case that to all intents and purposes gave birth to the field in which I ply my trade is one that for me exemplifies the role of the forensic anthropologist and how it fits into the judicial process. It should be an essential text for every forensic pathologist, anatomist, police officer, lawyer and judge as well as for every forensic anthropologist. This is a case that links Lancaster, where I currently live and work, with my homeland of Scotland. It shows police working alongside anatomists and features some groundbreaking detective and forensic skills that paved the way for my generation of scientists and investigators. It also underlines the need for us to be open-minded and receptive to all possibilities, and to be constantly refining our techniques and seeking new ways of reaching for the truth.

It is a story that gives an insight, too, into the lengths to which we must sometimes go in order to establish what happened to one of our fellow human beings and reminds us that at the end of every murder investigation lies the need for truth and justice. It demonstrates why we must know exactly what we can expect any body part found in isolation to be capable of revealing. It is perhaps an interesting exercise for us all to undertake. What could you find in your body that would help me to identify who you were and what life had thrown at you? Start at your head and work down to your toes, just as we have in this book, and you will be amazed at how many little things can be noted that, in combination, might create a likeness of you and your life that your family and friends would recognize.

At the centre of this case was Bukhtyar Rustomji Ratanji Hakim, who was born into a wealthy, middle-class French–Indian family in 1899. After qualifying in medicine and surgery he worked at a Bombay hospital and later for the Indian Medical Service. Seeking to widen his horizons, in 1926 he moved to London. He had big ambitions, but in a city full of aspiring medics, he found himself a rather uninspiring small fish in a very big pond. He relocated to Edinburgh, another highly respected seat of medicine and surgery, where he studied to become a Fellow of the Royal College of Surgeons, but failed the examinations on three separate occasions.

Feeling that his Indian name was holding him back, he changed it by deed poll to one he thought sounded much more British. And so it was as the debonair Dr Buck Ruxton that he met Isabella Kerr, the manageress of an Edinburgh restaurant. Bella, who was separated from her husband after a disastrous and short-lived marriage, saw this suave, exotic medic as her route to a better life.

Bella became pregnant, and to avoid a scandal they decamped to London, where they presented themselves as a married couple and she gave birth to their daughter. Ruxton once again found the capital a hard place in which to succeed. He finally decided that perhaps surgery was not for him, and that he might have a better chance of earning a reasonable living if he set up as a GP in an area where he would have less competition.

So, in 1930, the little family of three moved to Lancaster. This poor northern city, where there were not enough doctors to serve the population, was the perfect place for a new general practice to flourish. Property was cheap and Ruxton took out a loan to buy a Georgian townhouse at 2 Dalton Square, where he and Bella set up home and he opened his surgery.

Before long the practice was thriving. Ruxton was a popular GP and highly regarded by his patients. In particular he had an excellent reputation for his gynaecological skills at a time when mortality among pregnant women and infants was high. In this pre-NHS era, when all medicine and consulting had to be paid for, he was known to waive fees for his poorer patients who could not afford them.

With Buck's dapper appearance, medical expertise and kindly ways, and Bella's charm and social skills, the couple were quickly accepted by the local smart set. In five years they had added another two children to their family and on the surface life seemed good. Their house was comfortably furnished and they each had their own car, quite a status symbol in the 1930s. They had several domestics who cooked and cleaned for the family and a live-in maid, Mary Rogerson, who hailed from the nearby coastal town of Morecambe.

But beneath the glittering fairy story, all was not well between the couple. Bella was ambitious and headstrong. Not content to play the doctor's wife, she was set on having her own business and her own money. Buck wanted control, she wanted freedom and the outcomes of their often loud arguments were plain for all to see. Bella sported bruises around her neck and told the police, who were summoned on several occasions, that her husband was violent. She left him more than once, taking the children with her, because of his behaviour but always returned. The attitude then to domestic violence was that a man was the master of his family and however he chose to run his household, and manage his wife, was his business and nobody else's.

Everything about Bella's personality that had first beguiled Ruxton now became a source of fear and bitterness. He resented her independence and the money she spent on herself. Although no great beauty, Bella had a charisma that attracted other, younger men, and Buxton was insanely jealous. He became convinced that she had a lover and that she was going to leave him for good.

It all came to a head over the weekend of 14 September 1935. Bella had arranged a trip to Blackpool that Saturday night to visit two of her sisters who were living there and to see the world-famous illuminations. Ruxton was not happy about it. In the interests of a quiet life, rather than staying overnight as she had planned, she decided to drive back to Lancaster the same evening. But the fact that she did not arrive home until after 1 a.m. was confirmation, in his eyes, that she had been seeing another man.

As she entered the house in the early hours of the morning of Sunday 15 September, it is likely that Ruxton was waiting for her. He may have strangled her, as he had a history of this type of assault, or perhaps he lashed out with the poker. We will never know, because there were no witnesses. Whatever the sequence of events, Bella died. Maybe Mary Rogerson, the maid, hearing the commotion, came out on to the landing and met the same fate. However it happened, she, too, lost her life that morning. The amount of blood discovered subsequently on the stairs suggested that either one or both women may have been stabbed.

Did Ruxton set out to murder his common-law wife and their maid? Probably not, but they were dead none the less, and now he had to decide how to deal with it. Should he face the music and risk his career and reputation by owning up? Should he just pack his bags and run away? Or should he try to cover it up? He settled on the last option. He was undoubtedly an intelligent man but, inclined to arrogance, he had an over-inflated view of his own intelligence and may well have been a little contemptuous of the capabilities of the police. He would have to concoct a plausible story but, more pressingly, he would have to find a way of disposing of two bodies that were leaking blood and fluids all over his landing carpet.

Dismemberment must have seemed the logical solution to him. He had the necessary anatomical knowledge, he had studied forensic medicine and he had the surgical means. It is not enough, however, to have the expertise to cut up the bodies. You also have to be organized, to have an idea of where you are going to dump the parts and how you are going to handle the mess. Ruxton would need to dispose of the remains, clean up the house and come up with a cover story, all the while running his surgery and looking after the three children who were asleep in the house, minus the help of the maid.

He dragged the bodies across the landing into the bathroom, the preferred site for most dismemberments as it comes equipped with a body-shaped and appropriately sized vessel and a plumbing system to wash all the fluids away. He would have known he'd have to bleed the bodies, as he could not afford to leave a trail of blood through the house, and that he'd have to do it quickly, before the blood began to congeal and the task became more difficult. He would have to disfigure them to obscure their identities, aware that, in time, decomposition would do the rest of the job for him.

With the right equipment and the right skills, dismemberment does not actually take that long. He started with Bella, his common-law wife. Heaving her into the bath, he removed her clothes, skinned her torso and removed her breasts. He excised her larynx because he knew that the prominence of the Adam's apple was an indication of the difference between a man and a woman. He also removed her internal and external genitalia. He cut off her lips, ears, eyes, scalp and hair. He then removed her head. He sliced away her cheeks, pulled out her front teeth and all the others that had fillings or dental work which might identify her. He dismembered her whole pelvis and stripped her lower limbs of flesh because she had quite distinctive thick ankles. He removed the ends of her fingers, to prevent fingerprint comparison. He severed her major joints with skill and precision. He encountered only one significant setback: while getting rid of the evidence of a bunion on her right foot, his knife slipped and he cut his hand badly. This was going to slow him down and hinder him from doing such a thorough job on Mary.

By this point he was probably quite exhausted. His initial adrenaline rush would have hit the crash part of the cycle, he was wounded and his instruments would by now have been blunt and slippery. Although he removed many of the facial features that could have identified Mary, along with the skin from her thighs, to obliterate a birthmark, her hands and her feet he left alone. To what extent he dismembered her torso we do not know, as it was never found.

He made an excellent job of obscuring the identities of his victims. Perhaps too good a job, because in the process he left clues to his own. His clean disjointing of Mary's shoulder and hip was a clear indication that the person who had done this had an understanding of anatomy and possessed the necessary surgical skills. And the specific parts of the body he removed pointed to a high degree of knowledge of what was important to contemporary forensic identification.

Having dismembered the bodies, Ruxton locked the bathroom door, cleaned the landing carpet, and perhaps the walls, as best he could and changed his clothes, which would have been soaked in blood. Later that morning, he had breakfast with his children, visited his cleaning lady to say that she would not be required until the next day and dropped the children off with friends so that he could continue unobserved with the task awaiting him at home.

He wrapped the large body pieces in a mixture of old clothing and newspapers. He now had a big pile of body parcels, plus some clothing, identifiable excised body parts and remnants of tissue which he wanted to dispose of separately. He bought petrol and, over the course of several nights, he burned these in an old barrel in his back garden.

He invented a variety of stories to explain the absence of Bella and Mary, including the one he initially told Mary's parents—that she was pregnant and Bella had taken her away to procure an abortion for her. As abortions were illegal he hoped this would deter them from contacting the police. He managed to keep most people at bay for long enough to think through what he was going to do with all the parcels. There were comments from staff and patients about the condition of the house—the odd smell, and the disappearance of carpets—and on his dishevelled and exhausted appearance. He told some he was preparing the place for redecoration, others that Bella had left him again, attributing his state of disarray to stress and worry. His bandaged hand, he said, he had jammed in a door. But to be a really good liar, you have to be a consistent liar.

Ruxton realized he could not use his own vehicle to dispose of the bodies. He was too well known in the area. So he hired an unostentatious car, with a big boot, from a local company, and decided to drive north and dump the remains in Scotland, no doubt rationalizing that the police force across the border was unlikely to be in communication with their English counterparts. Having lived in Edinburgh, he knew the road well.

Early on Tuesday 17 September, taking his young son with him in the hired car, he drove towards the town of Moffat in the borders, a journey of over a hundred miles. With today's motorways and faster cars it can be done in just under two hours, but of course in 1935 it was a much longer expedition. A couple of miles north of Moffat in Dumfriesshire, he stopped on an old stone bridge across the Gardenholme Linn stream. There had been heavy rain and the burn was in full spate. He threw the contents of the boot over the parapet into the fast-flowing water.

At 12:25 p.m., a cyclist reported to the police that he had been knocked off his bike in Kendal by a speeding southbound motorist. He had taken a note of the registration number of the car, which was telephoned through to the next police station along the road, in Milnthorpe. A police officer was waiting there for Ruxton as he drove through, and stopped the car. The matter was recorded as a minor incident, since nobody was hurt, and Ruxton was allowed to continue on his way, having explained, ever the caring doctor, that the reason he had been driving so fast was that he had patients waiting to see him in Lancaster. But this was a huge mistake on Ruxton's part and he would have known it. His presence, and that of his hire car, in Cumbria, on the route back from the borders, had now been officially recorded, timed and dated.

Two days later he made the round trip again with the remainder of his cargo. This time he was more cautious and presumably disposed of the rest of the body parts, unseen, at various points in the Annan river and its tributaries.

By 25 September, Mary Rogerson's family had grown so concerned about her absence that they informed the police. Ruxton, her employer, was questioned and spun the constabulary one of his pre-prepared yarns.

On Sunday 29 September, fourteen days after Bella and Mary were murdered, a young woman out for a stroll near Moffat looked over the bridge into the Gardenholme Linn burn, as you do, and thought she saw a raised arm and a hand sticking out of the water. The local menfolk were fetched to take a closer look and found a bundle, caught against a boulder, containing a human head and upper limb. Police officers from the Dumfriesshire Constabulary were summoned and arrived swiftly on their bicycles.

A search of the burn, the surrounding streams and ravines and the Annan river brought forth dozens of body parts, including a second head. Some of them were wrapped in fabric or clothing and some in wet newspaper. These were the days, of course, before crime scene investigators, before DNA, before forensic photography or generators to light a scene through the night. The police officers were quick, thorough and efficient. They recovered everything they could find and made copious notes, paying admirable attention to detail. The body parts were taken to the mort house at the corner of the cemetery in Moffat, there to await the scrutiny of the doctors.

An inventory was taken the next day, in much the same way as we would do it now. So far they had two arms, two upper arm bones, two thigh bones, two leg bones, an upper trunk, the lower part of two legs, including feet, a pelvis, two disfigured heads and, in total, nearly seventy assorted pieces of human remains. The fabric and newspaper in which they had all been wrapped was removed, cleaned and carefully dried.

It was clear that these were not natural deaths and that the body parts belonged to a minimum of two people. It was also very apparent that the dismemberment had been performed with expertise. The police wondered if it may have been carried out by a doctor, or whether the whole thing might be a hoax, perhaps perpetrated by medical students dumping body parts from a dissecting room. What was not clear was whether the victims were local or whether the remains had been brought to Moffat from elsewhere. At this stage, then as now, establishing the possible identity of the two bodies was paramount if the killer was to be found and their story revealed.

It was obvious from the inventory that there were body parts still missing, and although a further search using dogs turned up a few more bits, what they had by no means comprised two complete bodies. An early assessment suggested that the victims might be an older man and a younger woman. This misdirection meant that nobody was looking for two missing women. There were no comparable people missing locally and so the investigation had to be widened. Easier to go north than south, especially as the remains had been found in Scotland, not England, and so the Glasgow police became involved, as did the anatomists and forensic doctors from the ancient universities of Glasgow and Edinburgh.

The principal anatomist was James Brash, a professor at Edinburgh university, and the two other academics were Sydney Smith, professor of forensic medicine at Edinburgh university, and John Glaister, professor of forensic medicine at Glasgow university. All three were esteemed academics with global reputations and, ironically, they would have been revered by Dr Ruxton. Professors Brash and Smith had probably taught him when he was studying for his surgeon's examinations.

They began to match the body parts, assigning them to individuals they named simply as Body 1 and Body 2. They suspected that the dismemberer had surgical or anatomical experience and they believed that the reason for the removal of some of the parts was to destroy something of interest that could have identified the bodies. And while they were aware that the purpose of the mutilation was to conceal sex and identity, they were still working on the premise that they had an older male and a younger female. Two tanks of embalming fluid were constructed, one for each body, and the assigned parts were placed in the fluid to halt further decomposition.

On 30 September, the newspapers carried reports of the grisly discovery at Moffat but the story still described the victims as a man and a woman. Ruxton must have felt mightily relieved. It was, though, those same newspapers that were to give the police the break they needed.

One of the newspapers used to wrap some of the remains was a copy of the *Sunday Graphic*, dated 15 September 1935 and carrying the serial number 1067. This provided the earliest possible date for the dumping of the bodies. Moreover, the newspaper was not only local to the Lancaster area but its circulation could be narrowed down even further: it was a limited "slip edition" distributed in small numbers only around Lancaster and Morecambe.

Their research into this newspaper brought the police down from Glasgow to Lancaster and Morecambe. There were no couples missing from round here, either, who fitted the description of the bodies, but there were two missing women. That must have been one of those real lightbulb moments—especially when the police learned that the husband of one of the women was a GP with a surgical background. They had been led down a blind alley for the first twelve days of the investigation: a classic illustration of why it is so important that the information given to the police in the early stages of a case is accurate.

The celebrated academics readily admitted that they could be wrong. Another vital lesson: not allowing our egos to take the investigation any further than necessary down that blind alley. By Sunday 13 October, Dr Ruxton had been charged with Mary's murder. Some of the items of clothing wrapped around the body parts had been recognized by her family.

Then came the first piece of brand-new forensic science. Whereas

Bella's fingertips had been cut off and were never recovered, Mary's were intact. The outer layer of epidermis on the hands of Body 1 was "degloved"—a condition known as "washerwoman hands"—from having been in the water for so long. However, her deeper, dermal fingerprints were visible. The fingerprint expert was able to retrieve the dermal prints from the corpse and compare them with epidermal fingerprints found in Mary's room at 2 Dalton Square, and indeed all around the house in areas that she helped to clean, including on glassware.

The dermal print, although finer and less clear-cut, has a similar ridge detail to the outer, epidermal print, and is just as valid for identification purposes. This was the first time that dermal fingerprinting played a role in a UK case and the first time such evidence would be admitted into court.

Rubber casts were taken of the hands and feet of Body 1, which fitted perfectly into Mary's gloves and shoes, but not Bella's. The skin removed from Mary's leg, which had held the birthmark, constituted negative evidence, speaking to the attempt to prevent identification rather than proving identity itself. If something is done to disfigure a specific part of a body, then it raises the question of what was there that someone was trying to hide.

Ruxton was charged with Mary's murder first because the evidence associated with her identity—sex, age, height, clothing, fingerprints, the absence of evidence in a critical area and the fit of her gloves and shoes to the hands and feet—was consistent with Body 1, described as a woman between the ages of eighteen and twenty-five with multiple blunt-force trauma injuries to the head. Body 2 was a female between the age of thirty-five and forty-five with five stab wounds to the chest and a fractured hyoid bone. Since Ruxton had been much more thorough in removing her identifying features, proving that this was Bella was going to be much more difficult.

The scientists turned their attention to photographs of the two women. The pictures in existence of Mary were not of good quality but there was an excellent image of Bella, wearing a diamond tiara. They decided to attempt to superimpose a photograph of a skull on to a photograph of a face, something that had never been tried before. This was the birth of the technique we were still using sixty years later when we assisted with the identification of two of the victims of the Monster of Terrazzo, as recounted in Chapter 2.

It was an inspired idea and an incredible feat of patience and perseverance on the part of the photographer. While it is true that the superimposition on to Mary's photograph was not convincing, in Bella's case it was a triumph, and the result remains an iconic forensic image to this day.

On 5 November, Ruxton was charged with the murder of his common-law wife. The Crown needed to secure sufficient evidence to get to the position of "beyond reasonable doubt," and it seems they felt they probably had enough, despite the fact that there were no witnesses to the assault, no murder weapon had ever been found and no confession was forthcoming. The case was almost entirely circumstantial, and therefore heavily reliant on some very new, untried and untested forensic techniques that prosecutors, police and scientists could only hope would be deemed admissible in court.

At trial, the Crown dropped the murder charge for Mary and the case went ahead based only on the evidence relating to Bella. This happens in court cases when a decision has to be made on the best strategy to be adopted to secure a conviction. Mary's family were understandably distraught that Ruxton would not face justice for taking the life of their daughter.

He entered a plea of not guilty and the Crown prepared eleven witnesses and 209 exhibits for the trial, which started on Monday 2 March 1936 at Manchester Court. Even though the primary investigating police force and the experts were Scottish, and the remains had been found in Scotland, the trial was held in England because that was where the offences had been committed. The case should have been heard at Lancaster Castle, but was moved to Manchester owing to concerns that there could be no fair trial in a small community where the accused was such a prominent figure.

It lasted eleven days, one of the longest murder trials on record in the English courts. The forensic evidence was admitted and, together with witness evidence from around the time of the murders and their aftermath, it constituted the majority of the testimony given. The final day, Friday 13 March, was certainly an unlucky one for Dr Ruxton. Having retired at 4 p.m., the jury returned just over an hour later with a unanimous guilty verdict and Mr Justice Sir John Singleton passed the death sentence. Buck Ruxton was to be taken from the court to Strangeways prison, where he would be hanged by the neck until dead.

Of course, Ruxton appealed, and his case was chaired by the lord chief justice, later Baron Hewart of Bury, on 27 April, but it was not upheld. Clemency was sought by the people of Lancaster with a petition of over 10,000 signatures, but this, too, was denied, and on 12 May 1936 the sentence was carried out. Ruxton was just thirty-six years old and left three orphaned children aged six, four and two. All sorts of distasteful things occur in the wake of cases as notorious as this one. Although the bodies of Bella and Mary were eventually buried, their skulls were retained by Edinburgh university. Charlatans offered messages from beyond the grave. Bawdy ditties were sung or recited in bars and playgrounds. The area where the bodies had been found became known locally as Ruxton's Dump. As for the house at Dalton Square, much of it had been dismantled for testing by the scientists, including the bath where the two women had been bled and dismembered. This ended up being used for many years as a horse trough by the Lancashire Mounted Police division in Hutton.

Would we go to the same lengths nowadays as the police and scientists of the 1930s? I would certainly hope so. Our predecessors left no stone unturned. As well as the careful preservation of evidence and the application of the remarkable innovations of superimposition and dermal fingerprinting, the case featured a Glasgow entomologist, Dr Alexander Mearns, who was brought on board to analyse pupae found on the remains to further narrow down the likely time of death—another technique that was in its infancy. This investigation had everything, and I would urge anyone interested in the full story to read Tom Wood's excellent book *Ruxton: The First Modern Murder*.

Today, of course, we would DNA sample the body parts to assign them to the correct individual and we would take matching DNA samples from Mary's parents and from Bella's children and her sisters. But we lose our core skills at our peril: we have no idea when we might need to call upon them. The latest techniques are not always able to give us answers.

We have become over-reliant on DNA and our testing methods are now so sensitive that we are starting to experience some challenges in the courtroom on the question of possible contamination. There are still things we don't know forensically about how DNA behaves. How it transfers on to different materials, for example, or for how long it stays there. We don't know how easy or difficult it is for it to be transferred from one surface to another and we have difficulty deconvoluting a mixed profile sample.

DNA evidence might be enough to prove identity but it may not be sufficient on its own to prove guilt or innocence in a court of law. So it is important to ensure that it is supported by as much corroborating evidence as possible. And when a case comes along where DNA cannot help we must rely on the scientific skill and knowledge of our various disciplines and the mutual acknowledgement that, when investigators and scientists work together as a team, we can achieve great things.

If we can reach a swift solution then of course we will, and sometimes the obvious answer will be the right one. The cases that stay in our minds, though, are not the easy ones but those that have been the hardest and required the most thought. What we always remember is that every body part we see belongs to a person who lived. Someone who had a mother, father, perhaps siblings or children, and friends and colleagues who cared about them.

As this journey through the human body has shown, the job of the forensic anthropologist is not to create a life story but to try to find and understand the story that has already been written large in its bones, muscles, skin, tendons, and in the very fibre of its being. We must be the bridge between whatever horrific, tragic or simply sad incident has brought about the end of a life and the handing back of the body in which it was lived to the people who will lay their loved one, and their story, to rest.

What it is not is rocket science, and although it may at times be portrayed as glamorous, it really isn't. It is a hard job that challenges you physically, intellectually and emotionally, but it is an honour and a privilege to play what is sometimes a very small part in the investigation process and to know that what you have done has made a difference to someone somewhere.

Before long, it will be my turn to pass the baton into the very capable hands of the generations coming up behind me who are better suited to the physically gruelling aspects of the job. I never pictured myself in the role of professional grandmother, but somehow it has sneaked up on me when I wasn't looking. When I saw myself on television recently in an interview I recorded with the criminologist David Wilson, I recognized, of course, many things that I already knew about myself. But by detaching the woman on the screen from "me," and viewing her as a different person, I think I can detect a lot more about her.

I see my mother and my father in me facially, but not in the way I talk. I don't have either their mannerisms or their accents. Like my father, I am capable of telling a story without necessarily answering the question I am being asked. My brain still runs faster than my mouth, and I can see myself thinking two steps ahead of what I am saying. I can tell when I am uncomfortable, from my body language and the tone of my voice, and when I am relaxed and feel myself to be on solid ground. I have two smiles and one of them does not reach my eyes. These are all characteristics I can recognize as part of who I am, but which no forensic anthropologist looking at my bones or my body when I am dead would ever be able to discern.

So we need to be realistic about what a stranger can tell from our bodies and just how valuable or otherwise that information may be in the identification of our mortal remains. I would hope that from whatever might be left of me a good forensic anthropologist would be able to determine that I was female, my age when I died, my height (5 ft 6 ins, or 1.67 m) and my red hair, if it is still red by then. If it isn't, they can always find that colouring in my genetic make-up, which would also tell them about my skin tone and whether or not I had freckles (I do). I would hope they could establish that my ancestry is Caucasian. A classic Celt.

They will see that I have no tattoos, no congenital abnormalities (of which I am aware), no deformities (though these may yet come), no modifications and as yet, please God, no amputations or major injuries. I have several scars from accidents, like the one under the ring on my right finger where I sliced it open with the lid of a tin of corned beef as a teenager. There is only one surgical scar to date, from an operation to reverse the gynaecological sterilization I had opted for. My pelvis might show signs of the three beautiful babies I brought into the world. My teeth would scream that I am a Scot with more fillings than teeth, and some have been extracted. I have no tonsils. I have the early stages of arthritis in my neck, back, hips and big toes. I broke my right collar bone years ago when I fell off my motorbike in the ice.

I have no surgical devices of any sort implanted in me. I have never been shot or stabbed. I have never taken any form of illegal substance (to my knowledge, at least) and toxicology would confirm that I am on no regular medication. All things considered, it is actually a pretty boring, run-of-the-mill body, so I can only apologize to the person who could have to sift through these bones looking for something, anything, that might be of interest.

I have said before that when I die I would like my body to be dissected in the anatomy department in Dundee university. I want to be embalmed using the Thiel method my department pioneered in the UK. The very ordinariness of my remains will make me an excellent silent teacher. I would prefer my students to be scientists, please, rather than medics or dentists. Science students learn about anatomy in much more detail, because their curriculum devotes more time to the subject. When they have finished with me, I would like all my bones to be brought together and boiled down to get rid of all the fat inside, and then to be rearticulated as a teaching skeleton so that I can hang in the dissecting room I helped to design and continue to teach for the rest of my death.

I see it as such a waste to go up in smoke or to be buried uselessly in the soil. And what could be more fitting, for an anatomist and forensic anthropologist, than to want to be an articulated skeleton when she grows up?

# Acknowledgements

Acknowledgements are tricky things to write when there are so many people to thank. The production of a book is a genuine team effort and the writer is just one member of the cast. I hope it goes without saying that my husband and my beautiful daughters are top of the list. They have tolerated my reclusive retreating into my attic for hours, days and weeks at a time. They have thrown in food and gallons of tea from the door and endured frequent bouts of outwardly directed frustration and doubt. Without them, I have no purpose.

Then there is my second family, who have looked after me so well for a number of years and for whom I struggle to find words that express how much I adore them. To the lovely and marvellously bonkers Susanna Wadeson and the unflappable and smooth-talking Michael Alcock—that tea and cake with you both in the Wellcome cafeteria has so much to answer for. How you persuaded me to do this I will never know, but thank you.

To Caroline North McIlvanney, who wonderfully labelled herself as my "éminence grise." You are never in the shadows, lovely lady, but are writ large across every page and for that I remain in awe of your talents and am eternally grateful.

Steph Duncan stepped into the story for a while and although it was a cameo appearance, it was fun. Thank you for keeping an expert hand on the tiller and steering us in the right direction.

And then there is a third family. Creative and accomplished people I may only have met once or twice, but who work away in the background, doing what they do so magnificently to make this all happen: Kate Samano, Sharika Teelwah, Katrina Whone, Cat Hillerton, Tabitha Pelly, Emma Burton and all of the Transworld team. I am also indebted to the indomitable legend that is Patsy Irwin for her continued sound guidance and the incredibly talented Richard Shailer for his artistic brilliance. I thank you all from the bottom of my heart.

I want to pay a heartfelt tribute to one other very small group: the anatomists and the forensic anthropologists who have taught me so much and with whom I have shared so many adventures over the years.

My early teachers of anatomy are gone now but they instilled a passion that has granted me a fulfilling career. My belated thanks are due to Professor John Clegg and Professor Michael Day for the faith they showed in me.

To Professor Louise Scheuer and Professor Roger Soames: we have had some amazing times together and I have learned so much from you, even if I didn't always listen. Soz!

Finally, to my wingman. The forensic anthropologist who has a special place in my heart and with whom I have shared many experiences so ridiculous that we could never, ever publish them. Everyone should be lucky enough to have a Lucina Hackman in their life and I am grateful that I have been blessed to know and work with the original. Friend, colleague and partner in crime.

# Index

# About the Author

**Professor Dame Sue Black** is one of the world's leading anatomists and forensic anthropologists. She was the lead anthropologist for the British Forensic Team's work in the war crimes investigations in Kosovo. She was one of the first forensic scientists to travel to Thailand following the Indian Ocean tsunami to provide assistance in identifying the dead. Sue is a familiar face in the media, where documentaries have been filmed about her work, and she led the highly successful BBC Two series *History Cold Case*. She is the author of the critically acclaimed *Sunday Times* bestseller *All That Remains*.

Sue was appointed Dame Commander of the Order of the British Empire in the 2016 Queen's Birthday Honours for services to forensic anthropology. She is President of St John's College, Oxford.